Sietze van Buuren

Modeling and simulation of porous journal bearings in multibody systems

Karlsruher Institut für Technologie
Schriftenreihe des Instituts für Technische Mechanik
Band 21

Eine Übersicht über alle bisher in dieser Schriftenreihe
erschienene Bände finden Sie am Ende des Buchs.

Modeling and simulation of porous journal bearings in multibody systems

by
Sietze van Buuren

Dissertation, Karlsruher Institut für Technologie (KIT)
Fakultät für Maschinenbau
Tag der mündlichen Prüfung: 5. April 2013

Impressum

Karlsruher Institut für Technologie (KIT)
KIT Scientific Publishing
Straße am Forum 2
D-76131 Karlsruhe
www.ksp.kit.edu

KIT – Universität des Landes Baden-Württemberg und
nationales Forschungszentrum in der Helmholtz-Gemeinschaft

KIT Scientific Publishing 2013
Print on Demand

ISSN 1614-3914
ISBN 978-3-7315-0084-1

Modeling and simulation of porous journal bearings in multibody systems

Zur Erlangung des akademischen Grades

Doktor der Ingenieurwissenschaften

der Fakultät für Maschinenbau

des Karlsruher Instituts für Technologie (KIT),

genehmigte

Dissertation

von

M.Sc. Sietze W. van Buuren

aus Elsloo (Friesland) in den Niederlanden

Tag der mündlichen Prüfung: 5. April 2013

Hauptreferent: Prof. Dr.-Ing. Wolfgang Seemann

Korreferent: Prof. Dr.-Ing. Bernhard Schweizer

Foreword

This thesis is the result of three years of hard work on the subject of plain journal bearings in the field of multibody dynamics. During this time modeling approaches were developed that fill the gap between very simple and fast solutions and accurate but slow numerical solutions for the plain journal bearing problem. I am sure the results will be a valuable contribution to this field of research and I am looking forward to see where the developed modeling approach will be of use to scientists and engineers in industry *and* academia. I already know that the developed model is being applied to some multibody models of actual products and that it contributed to the predictions which can be used to improve these products and shorten their development process.

This thesis was started as one of the cornerstones of a cooperation between the department of contact dynamics (CR/ARU) belonging to the corporate research business unit of the company Robert Bosch GmbH and the Institute of Engineering Mechanics (Institut für Technische Mechanik, ITM), Department of Mechanical Engineering, Karlsruhe Institute of Technology (KIT), formerly known as Universität Karlsruhe (TH). This cooperation was centered around Dr.-Ing. Hartmut Hetzler and Prof. Dr.-Ing. Wolfgang Seemann who represented the university in this collaboration and also supervised this work. On the other hand, the company was represented by Dr. Markus Hinterkausen for the supervision. I would like to thank these people, without them the current level of quality would not have been reached. All those long hours of discussion concerning my work really motivated and helped me to push even further and extend my boundaries. For this I am very thankful.

Very important for the current work was also Prof. Dr.-Ing. Bernhard Schweizer due to his valuable input and feedback as a second reviewer, for which I am very grateful. Moreover, I would like to thank Prof. Dr.-Ing. Carsten Proppe, Prof. Dr.-Ing. Alexander Fidlin, Prof. Dr.-Ing. Jörg Wauer, Prof. Dr.-Ing. Walter Wedig and Prof. Dr.-Ing. Jens Wittenburg for their remarks and comments on my work.

Also, I would like to thank my colleagues at the department of contact dynamics at Robert Bosch GmbH and of course I also would like to thank the colleagues at the Institute of Engineering Mechanics, Department of Mechanical Engineering at the KIT. I really enjoyed the yearly seminars of the institute and appreciated their valuable input and questions. Especially, I enjoyed the *BookCamp* we had, where I learned an incredible amount on the subject of non-linear dynamics. Moreover, I would like to thank Jan Röper for his work on the influence of elastic material deformation in porous bearings.

Last but not least, I have to thank Claudia Schubert, my beloved girlfriend, for always being there for me and supporting me through every stage of my Ph.D.

Sietze van Buuren
Gerlingen, August 18, 2013

Abstract

The plain journal bearing is a widely used machine element to mount rotating machinery. A specific cost-efficient type of plain journal bearing is the porous journal bearing, which possesses a pervious bush that is filled with lubricant giving it self-lubricating properties. Increasingly, the operating conditions of porous bearings are extended to higher loads (often with changing direction), intermittent motion and higher rotor speeds. Because the current work is concerned with modeling plain journal bearings in multibody systems the interaction of plain journal bearings with surrounding machine elements is also of interest. Instead of steady state models that are typically utilized to analyze the behavior of plain journal bearings, fully dynamical models are needed to investigate the bearing's behavior under these operating conditions. To solve the plain journal bearing problem the differential equations governing the flow in the fluid film and the porous bush need to be solved simultaneously in order to obtain the reaction forces and moments on the surrounding system. A particular simple approach is to use simplified analytical approximations, which are fast but neither are valid for the general case nor for porous journal bearings. Numerical discretization methods on the other hand are valid for the general case and can be easily extended to include several physical phenomena, but are too time-consuming for the current application. To overcome these issues this work presents a semi-analytical approach that is based on the mesh-free Galerkin method, which yields a discretized description with only few unknowns. It is, however, able to account for the relevant physical phenomena such as the influence of the porous bush, surface roughness and misalignment of the journal's axis with respect to the bearing's axis. The proposed bearing model moreover enables symbolic calculation of stiffness and damping coefficients that can be used to construct the Jacobian matrix efficiently. This matrix is employed to improve the robustness and performance of the time integration and also may be used to analyze the stability of steady state solutions.

An important effect in journal bearings is cavitation or fluid film rupture in the diverging film thickness range. It is shown that the Gümbel cavitation algorithm, which is used for the proposed bearing model, is a good approximation of numerical discretization methods with the more realistic Reynolds cavitation algorithm. The model is validated for low and high load cases by comparison of the friction coefficient with experimental results in the hydrodynamic regime. It is observed that for increasing load the pervious nature of the porous journal bearings has less influence and eventually is better described using solid journal bearing models.

The hydrodynamical load capacity of solid journal bearings approaches infinity when the minimum film thickness goes to zero, whereas for porous journal bearings the load capacity is finite when the minimum film thickness approaches zero. The same phenomenon is observed for the stiffness and damping coefficients. It is shown that surface roughness does not significantly influence the hydrodynamical pressure but becomes influential through asperity contacts.

Finally, using the proposed bearing model stability and bifurcation analyses of elementary rotor-bearing systems are carried out. It is found that for balanced rotor-bearing systems the critical rotor speed decreases for increasing permeability but increases for increased applied load. The nonlinear dynamical behavior of unbalanced rotor-bearing systems is analyzed using path-following software that can follow the periodic solutions of the rotor-bearing systems for varying rotor speed and identify so-called bifurcation points. These are studied for varying applied load, permeability and rotor unbalance. The influence of misalignment in plain journal bearings is demonstrated by studying the nonlinear dynamics of rotor-bearing systems with a flexible rotor.

Finally, an application of the proposed bearing model to a multibody system containing porous journal bearings is presented.

Therewith, it has been shown that the proposed bearing model is a viable approach to study the dynamic behavior of (multibody) systems containing plain journal bearings, even beyond the onset of (local) instability.

Zusammenfassung

Gleitlager sind weitverbreitete Maschinenelemente, um rotierende Maschinen zu lagern. Eine häufig verwendete und kosteneffiziente Gleitlagerart ist das Sintergleitlager: es verfügt über eine poröse Buchse, die aufgrund ihrer Schmiermittelfüllung selbstschmierende Eigenschaften hat. Ständig erweiterte Betriebsbedingungen stellen zunehmend höhere Anforderungen, weil sie unter immer höheren Belastungen, wechselnden Richtungen und höheren Drehgeschwindigkeiten eingesetzt werden. Da sich die vorliegende Arbeit mit der Modellierung der Dynamik von Gleitlagern und Sintergleitlagern in Mehrkörpersystemen beschäftigt, werden auch Interaktionen von Gleitlagern mit umgebenden Maschinenelementen betrachtet. Anstelle von Modellen des stationären Zustands, welche typischerweise zur Analyse des Verhaltens von Gleitlagern angewendet werden, wird deswegen mit Hilfe von vollständig dynamischen Modellen das Verhalten von Gleitlagern und Sintergleitlagern unter entsprechenden Betriebsbedingungen untersucht.

Um eine Lösung für dieses Gleitlagerproblem zu finden, müssen die Differenzialgleichungen für den Schmierfilm und die poröse Buchse simultan gelöst werden: dies liefert letztlich die Reaktionskräfte und -momente auf das umgebende System. Ein besonders einfacher Ansatz ist die Anwendung analytischer Näherungen, die schnell sind, jedoch weder das allgemeine Gleitlager noch das Sintergleitlager korrekt abbilden können. Numerische Diskretisierungsverfahren hingegen liefern bei Gleitlagern im Allgemeinen bessere Ergebnisse und können einfach erweitert werden, um weitere physikalische Phänomene in die Betrachtung einzubeziehen. Jedoch sind diese Verfahren sehr zeitaufwendig für die vorliegende Anwendung. In dieser Arbeit werden die genannten Probleme mit Hilfe eines semi-analytischen Ansatzes gelöst, der auf der Galerkin-Methode beruht. Dieser Ansatz liefert eine Beschreibung mit wenigen Unbekannten, der zeiteffizient gelöst werden kann. Dennoch werden die relevanten physikalischen Phänomene wie der Einfluss der porösen Buchse, Oberflächenrauigkeit und Verkippung der Wellenachse zur Lagerachse berücksichtigt. Die vorgeschlagene Methode bietet außerdem die Möglichkeit, Steifigkeits- und Dämpfungskoeffizienten ohne numerische Differenzierung zu berechnen, um sie für die effiziente Erstellung der Jakobi-Matrix zu verwenden. Diese Matrix kann zur Verbesserung der Robustheit und Leistung der numerischen Zeitintegration eingesetzt werden. Weiterhin kann sie zur Berechnung der (lokalen) Stabilitätsgrenze eingesetzt werden.

Ein weiterer wichtiger Effekt in Gleitlagern ist die Kavitation bzw. das Abreißen des Schmierfilms im divergierenden Anteil des Schmierspaltes. Es hat sich gezeigt, dass der hier verwendete Gümbel-Kavitation-Algorithmus, eine gute Annäherung an Ergebnisse liefert, die mit numerischen Diskretisierungsverfahren unter Verwendung des realistischeren Reynolds-

Kavitations-Algorithmus berechnet wurden. Das Modell wurde validiert für Fälle mit niedrigen und hohen Belastungen, wobei die Reibungskoeffizienten mit experimentellen Ergebnissen im hydrodynamischen Bereich verglichen wurden. Es wurde beobachtet, dass der Einfluss der Durchlässigkeit von Sintergleitlagern bei steigender Tragkraft sinkt und deshalb effizienter unter Verwendung von Modellen für klassische Gleitlager mit nicht poröser Buchse beschrieben werden kann.

Die hydrodynamische Tragkraft von Gleitlagern mit nicht poröser Buchse geht gegen Unendlich, wenn sich das Minimum der Schmierfilmhöhe der Null nähert. Im Gegensatz hierzu ist die Tragkraft von Sintergleitlagern stets endlich, selbst wenn das Minimum der Schmierfilmhöhe Null beträgt. Das gleiche Phänomen lässt sich auch für Steifigkeits- und Dämpfungskoeffizienten beobachten. Weiter hat sich herausgestellt, dass sich die Oberflächenrauigkeit zwar nicht signifikant auf den hydrodynamischen Druck auswirkt, aber durch Kontakte von Rauigkeitsspitzen an Einfluss gewinnt.

Abschließend wurden Stabilitäts- und Bifurkationsuntersuchungen von einfachen Rotor-Lager-Systemen durchgeführt. Bei unwuchtfreien Rotor-Lager-Systemen nimmt die kritische Drehgeschwingigkeit der Welle bei steigender Permeabilität ab, aber nimmt bei steigender aufgebrachter Belastung zu. Das nichtlineare dynamische Verhalten von Rotor-Lager-Systemen mit Unwucht wurde mit Hilfe einer Pfadverfolgungssoftware untersucht, welche die periodische Lösung der Rotor-Lager-Systeme bei variierender Geschwindigkeit verfolgen und Bifurkationspunkte identifizieren kann. Hierbei wurde insbesondere der Einfluss von Belastung, Permeabilität und Rotorunwucht beleuchtet. Der Einfluss von Verkippung in den Gleitlagern wurde am Beispiel eines Rotorsystems mit flexiblem Rotor untersucht.

Zum Schluss wird eine Anwendung der vorgeschlagenen Methode auf ein Mehrkörpersystem mit Sintergleitlagern vorgestellt.

Hiermit ist gezeigt, dass das vorgeschlagene Modell eine praktikable Methode ist, das dynamische Verhalten von (Mehrkörper-) Systemen mit Gleitlagern zu untersuchen. Es kann sogar angewandt werden, wenn das System nicht mehr lokal stabil ist.

Contents

List of Figures

Nomenclature

Abbreviations & acronyms

BLAS   Basic Linear Algebra Subprogram

DOF   Degree(s) of Freedom

FD   Finite Difference

FEM   Finite Element Method

JFO   Jakobsson-Floberg-Olsson

LAPACK   Linear Algebra PACKage

SOR   Successive-Over-Relaxation

Greek symbols

α Angle between the eccentricity vector and the pure squeeze velocity vector .... [rad]

$\tilde{\alpha}$ Beavers-Joseph's slip-coefficient .... [-]

β Ratio between the shaft and bearing length, $\beta = \ell/L$ .... [-]

$\hat{\beta}$ Mean radius of the asperity peak curvatures .... [m]

δ_i Deviation of a rough surface from the datum, i refers to the surface .... [m]

$\boldsymbol{\varepsilon}$ Dimensionless eccentricity vector, $\boldsymbol{\varepsilon} = \boldsymbol{e}/c$ .... [-]

ε Norm of the dimensionless eccentricity vector, $\varepsilon = \|\boldsymbol{\varepsilon}\|$ .... [-]

$\bar{\varepsilon}$ Dimensionless journal axis projection on the mid-plane .... [-]

$\varepsilon_{ij}^{\text{el}}$ Components of the strain tensor .... [-]

η Dynamic fluid viscosity .... [Pa s]

$\hat{\eta}$	Density of asperity peaks on a rough surface	$[\mathrm{m}^{-2}]$
$\tilde{\eta}$	Effective viscosity, as defined in Brinkman's model	$[\text{-}]$
γ	Attitude angle	$[\mathrm{rad}]$
$\tilde{\gamma}$	Peklenik factor	$[\text{-}]$
ι	Ratio between the bearing length and the radial clearance, $\iota = L/c$	$[\text{-}]$
κ	Ratio between the inner bearing radius and the bearing length, $\kappa = R_i/L$	$[\text{-}]$
κ_o	Ratio between the outer bearing radius and the bearing length, $\kappa_o = R_o/L$	$[\text{-}]$
μ	Friction coefficient	$[\text{-}]$
μ_c	Friction coefficient parameter	$[\text{-}]$
ν_i	Poisson's ratio of contact body i	$[\text{-}]$
Ω_i	Angular frequency of the journal ($i = 2$) and the bearing ($i = 1$)	$[\mathrm{rad\,s^{-1}}]$
Ω^*	Dimensionless sum angular frequency, $\Omega^* = (\Omega_1 + \Omega_2)/\Omega_0$	$[\text{-}]$
Ω_s^*	Dimensionless sliding angular frequency, $\Omega_s^* = (\Omega_2 - \Omega_1)/\Omega_0$	$[\text{-}]$
$\boldsymbol{\phi}$	Vector of rotation angles between the journal and the bearing	$[\mathrm{rad}]$
ϕ_i	Rotation angle between the journal and the bearing, $i = x, y, z$	$[\mathrm{rad}]$
$\phi_{x,\theta}^f, \phi_z^f$	Pressure flow factors in circumferential (x or θ) and axial (z) direction	$[\text{-}]$
$\boldsymbol{\Phi}$	Monodromy matrix	
ϕ_s^f	Shear flow factor	$[\text{-}]$
Ψ	Dimensionless permeability, $\Psi = KR_i/c^3$	$[\text{-}]$
ψ	Angle between the eccentricity vector and the journal axis projection the mid-plane $[\mathrm{rad}]$	
ρ	Fluid density	$[\mathrm{kg\,m^{-3}}]$
σ_e	Composite standard deviation of the asperity peaks, $\sigma_e = \sqrt{\sigma_{e,1}^2 + \sigma_{e,2}^2}$	$[\mathrm{m}]$
$\sigma_{e,i}$	Standard deviation of the asperity peaks, i refers to the surface	$[\mathrm{m}]$
σ_h	Composite standard deviation of the roughness profile, $\sigma_h = \sqrt{\sigma_{h,1}^2 + \sigma_{h,2}^2}$	$[\mathrm{m}]$
$\sigma_{h,i}$	Standard deviation of the roughness profile, i refers to the surface	$[\mathrm{m}]$
σ_{ij}^{el}	Components of the stress tensor	$[\mathrm{Pa}]$
τ	Dimensionless time, $\tau = \Omega_0 t$	$[\text{-}]$

θ	Circumferential angle	[rad]
ξ^{m}	Ratio between the shaft and disk mass (for flexible shafts)	[-]

Roman symbols

a Rotor unbalance .... [m]

b_{ij}^{*} Dimensionless damping coefficient, $i,j = x,y$ .... [-]

c Radial clearance .... [m]

c_{GW}^{i} Asperity contact coefficient, $i = sr, rr$ .... [-]

D Journal diameter .... [m]

$\tilde{d}$ Mean conduit diameter of the porous medium .... [m]

e Eccentricity vector .... [m]

$\bar{e}$ Journal axis projection on the mid-plane .... [m]

E Composite elastic modulus of the contact bodies, $1/E = (1 - v_1^2)/E_1 + (1 - v_2^2)/E_2$ [Pa]

E_i Elastic modulus of contact body i .... [Pa]

E^{w} Elastic modulus of the flexible shaft .... [Pa]

$\bar{f}$ Porosity .... [-]

F_k Total ($k =$ t), hydrodynamical ($k =$ h), asperity contact ($k =$ e) or externally applied ($k =$ a) force .... [N]

$F_{(k,j)}$ Force (component), where $j = \varepsilon, \gamma, x, y$ .... [N]

$F_{(k,j)}^{*}$ Dimensionless force (component), $F_{(k,j)}^{*} = (c/R_i)^2 F_{(k,j)}/\eta \Omega_0 LD$ .... [-]

$F_{\mu,k}^{*}$ Dimensionless hydrodynamic ($k =$ h) and asperity contact ($k =$ e) friction force .. [-]

F_n Probability density function of the surface roughness contact, where $n = 3/2, 5/2$ [-]

g Gravitational acceleration vector .... [m s^{-2}]

H Fluid film thickness function .... [m]

h Dimensionless fluid film thickness function, $h = H/c$ .... [-]

H_T Fluid film thickness function including random roughness ...... [m]

$\mathscr{I}$ Moment of inertia tensor .... [kg m^2]

I_i	Moment of inertia about axis i ($i = x, y, z$)	$[\text{kg m}^2]$
I^{w}	Second moment area of the flexible shaft	$[\text{kg m}^2]$
$\boldsymbol{J}$	Jacobian matrix	
$\mathscr{K}$	Permeability tensor	$[\text{m}^2]$
K	Permeability	$[\text{m}^2]$
k_{ij}^*	Dimensionless stiffness coefficient, $i, j = x, y$	$[\text{-}]$
L	Bearing length	$[\text{m}]$
ℓ	Shaft length	$[\text{m}]$
$\mathscr{L}$	Elasticity tensor	$[\text{Pa}]$
M	Number of ansatz functions in axial direction	$[\text{-}]$
m	Rotor mass	$[\text{kg}]$
M_k	Total ($k = \text{t}$), hydrodynamical ($k = \text{h}$) or asperity contact ($k = \text{e}$) moment	$[\text{N m}]$
$M_{k,j}$	Moment component where $j = \varepsilon, \gamma, x, y$	$[\text{N m}]$
$M_{(k,j)}^*$	Dimensionless moment (component), $M_{(k,j)}^* = (c/R_i)^2 M_{(k,j)} / \eta \Omega_0 L^2 D$	$[\text{-}]$
N	Number of ansatz functions in circumferential direction	$[\text{-}]$
$\tilde{N}$	Characteristic size scale of the pore structure of the porous medium	$[\text{-}]$
p_{h}	Pressure in the fluid film	$[\text{Pa}]$
p_{h}^*	Dimensionless pressure in the fluid film, $p_{\text{h}}^* = (c/R_i)^2 p_{\text{h}} / 6 \eta \Omega_0$	$[\text{-}]$
$\widehat{p}_{\text{h}}$	Fluid pressure in the porous medium	$[\text{Pa}]$
$\widehat{p}_{\text{h}}^*$	Dimensionless fluid pressure in the porous medium, $\widehat{p}_{\text{h}}^* = (c/R_i)^2 \widehat{p}_{\text{h}} / 6 \eta \Omega_0$	$[\text{-}]$
p_{e}	Asperity contact pressure	$[\text{Pa}]$
$\boldsymbol{q}$	Fluid velocity vector	$[\text{m s}^{-1}]$
$\widehat{\boldsymbol{q}}$	Fluid velocity vector inside the porous medium	$[\text{m s}^{-1}]$
q_i	Fluid velocity in the direction i, where $i = x, y, z$	$[\text{m s}^{-1}]$
r	Radial coordinate	$[\text{m}]$
r^*	Dimensionless radial coordinate, $r^* = r/R_i$	$[\text{-}]$
R	Journal radius	$[\text{m}]$
R_i	Inner bearing radius	$[\text{m}]$

R_o	Outer bearing radius	[m]
s	Stiffness parameter related to the Euler-Bernoulli beam element	[Pa m]
s^*	Dimensionless stiffness parameter, $s^* = c^3/R_i^2 s/\eta\Omega_0 LD$	[-]
t	Time	[s]
U_i	Velocity of the contact surface i in x-direction	[m s^{-1}]
$\boldsymbol{u}$	State space vector	
U_s	Sliding velocity, $U_s = U_1 - U_2$	[m s^{-1}]
V_i	Velocity of the contact surface i in y-direction	[m s^{-1}]
$\boldsymbol{v}_s$	Pure squeeze velocity vector	[m s^{-1}]
$\boldsymbol{v}_s^*$	Dimensionless pure squeeze velocity vector, $\boldsymbol{v}_s^* = \boldsymbol{v}_s/c\Omega_0$	[-]
v_s^*	Norm of the dimensionless pure squeeze velocity vector, $v_s^* = \|\boldsymbol{v}_s^*\|$	[-]
$\mathcal{W}_i^*$	Dimensionless impedance vector component with direction i	[-]
W_i	Velocity of the contact surface i in z-direction	[m s^{-1}]
x	Horizontal coordinate	[m]
$\tilde{x}$	Horizontal coordinate in the pure squeeze coordinate system	[m]
y	Vertical coordinate	[m]
$\tilde{y}$	Vertical coordinate in the pure squeeze coordinate system	[m]
z	Axial coordinate	[m]
z^*	Dimensionless axial coordinate, $z^* = z/L$	[-]

Chapter 1

Introduction

1.1 Motivation

The plain journal bearing has been used extensively during the past centuries to mount rotating machinery. Alongside a steady increase in application, knowledge about these machine elements also has grown gradually. More recently, the application of a special type of plain journal bearing – the porous journal bearing – has become popular due to its advantageous self-lubricating properties and low production costs.

This thesis focuses on modeling the dynamical behavior of plain journal bearings with cylindrical bushings fabricated from both impervious and pervious materials. The classical solid journal bearing has been covered extensively in literature and in this thesis it will mostly serve as a foundation to understand the behavior of porous journal bearings. The models proposed in this thesis can also be applied to solid journal bearings, however, simply by assuming that the porous material is impervious.

Compared to solid journal bearings, the modeling of porous journal bearings is still a relatively new topic in literature and many questions on the behavior of porous journal bearings remain unanswered. This is especially the case when dynamical effects are additionally taken into consideration. Porous bearings usually are restricted to low rotor speeds and relatively low loads, but due to higher cost reduction demands manufacturers have sought to replace more expensive types of bearings with porous journal bearings. This results in higher operating condition requirements such as higher loads, changing load direction and higher rotor speeds. Effects such as mixed lubrication, bearing misalignment and the nonlinear dynamical properties of the fluid film are influential phenomena in these regimes. It can be concluded that there is a need for

Figure 1.1: Porous metal bearings

simulation methods that account for these conditions. With the increase of calculation power, prediction of the dynamical behavior of products containing e.g. plain journal bearings has become more realistic. This has made possible the simulation of so-called multibody systems, which finds a wide range of applications in industry. The simulation of plain journal bearings in multibody system requires special attention. Large systems containing several bearings, require fast and robust models for time-efficient simulations.

Most of the current models are either too detailed and not applicable due to long calculation times and high memory allocation demands or are too simple and therefore fail to describe the essential physical behavior. This work proposes a semi-analytical bearing model, which is based on a mesh-free Galerkin approach. The model yields a very compact and small discretization of the problem and thus allows for a very efficient solution scheme, however, it still captures the essential physical behavior.

These fast but accurate models now make it possible to study the behavior of nonlinear dynamical systems containing plain journal bearings. In this work this is demonstrated firstly by analyzing elementary rotor-bearing systems containing porous and solid journal bearings to understand their fundamental stability and bifurcation behavior. Finally, a potential application of the method to a multibody system using the commercial multibody system software ADAMS[1] is presented.

---

[1] © MSC software, for more information see http://www.mscsoftware.com

1.2 Literature survey

This section features an overview of the literature on plain journal bearings with an emphasis on the porous journal bearing. To this end, the classical solid journal bearing (i.e. impervious plain journal bearing) and the porous journal bearing[2] shall briefly be introduced. Subsequently literature on several topics concerning both types of journal bearings will be discussed.

Hydrodynamic lubrication of thin films

A journal bearing consists of a journal (or shaft) which rotates and moves inside the bearing (or bush) and is typically lubricated by oil, however many other fluids or even gases can be used here. Although the journal bearing has been used for many centuries now, one of the first major theoretical improvements was made by Reynolds who simplified the Navier-Stokes equations for thin fluid films. Since the surfaces of the journal and the bearing have nearly identical shapes the contact can be considered conformal. This assumption combined with several other simplifications results in the Reynolds equation, which is a differential equation for the fluid pressure [135, 149]. This governing equation is used to describe the hydrodynamical pressure in the fluid film of a plain journal bearing and therewith also its load capacity. The total load capacity of plain journal bearings is also influenced by several other phenomena, which will be introduced throughout this literature survey.

Porous journal bearing

Since the first hydrodynamic model for a porous journal bearing was published by Morgan and Cameron [106] many researchers have been working on this topic. The distinguishing property of a porous bearing is its porous bush. This bush usually is made from sintered bronze or iron. [21, 106] describe the fabrication process of porous bearings in detail. Porous bearings are maintenance free, offer high precision, produce low noise levels and possess low friction values, all for relatively low production costs. Despite these good properties porous bearings are not able to carry high loads and here solid journal bearings or ball bearings are the preferred alternative [6].

Due to its porosity the porous bush can absorb lubricant and sustain a lubricant flow (see figure 1.2). Typically fluid is pressed into the bush where a fluid film has formed and exits into the clearance outside the fluid film extent. Pressure formation still is very similar to classical

---

[2]To avoid misconception, unless stated otherwise throughout this thesis, the solid journal bearing will refer to the plain journal bearing with an impervious bush.

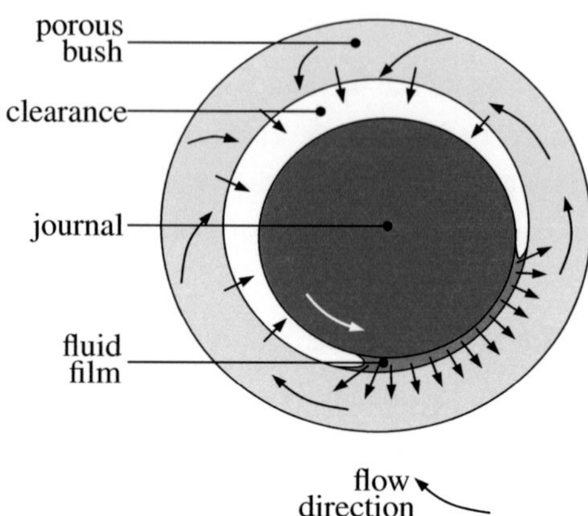

Figure 1.2: Lubricant flow in a porous bearing with respect to the fluid film.

journal bearings: due to an eccentric displacement and/or motion of the journal pressure will typically build up where the clearance converges and it will rapidly decrease in the diverging part of the clearance where pressure is for the most part atmospheric or depends on the surrounding pressure. However, lubricant circulation in the clearance and the bush as well as fluid exchange through the porous-fluid interface have an important impact on the fluid film extent and influence porous bearing characteristics significantly.

An important parameter of the porous material is permeability, which is the ability of the material to transmit fluid. For low Reynolds numbers flow through a porous material can be described with Darcy's law [34][3]. Darcy's law relates the fluid's flow velocity to its pressure and will simplify to the Laplace equation of pressure, when the permeability of the porous material and viscosity of the fluid are assumed isotropic and constant.

Cavitation

Cavitation is an important effect in journal bearings and should be taken into account for both porous and solid journal bearings. Additional conditions are needed to model the effect of cavitation together with the conventional Reynolds equation. Cavitation takes place in the diverging part of the clearance where the fluid film has ruptured and a mixture of gas and fluid remains [41, 42]. If no cavitation condition is prescribed, the Reynolds equation will solve to the so-called SOMMERFELD solution. The pressure will then become negative in the diverging part of

---

[3]Actually, for higher Reynolds numbers the Forchheimer equation is the correct relationship [117], but under normal operating conditions Reynolds number stays low in porous bearings and it suffices to use Darcy's law [79].

4

the clearance, which is physically unrealistic[4]. The most simple way to incorporate cavitation is to only take into account the positive part of the SOMMERFELD solution, which is called the HALF-SOMMERFELD or GÜMBEL solution. This approach is still applied nowadays to plain journal bearings, mostly in studies that take into account the dynamical behavior for solid journal bearing e.g. [16, 25, 152] and for porous journal bearings [31, 33, 97, 106, 108, 131].

A more sophisticated way is the REYNOLDS or SWIFT-STIEBER boundary condition, which requires the derivative of pressure to vanish at the rupture boundary of the fluid film, which gives rise to a free-boundary value problem since the position of the boundary is a priori unknown [145, 146]. It is more accurate than the GÜMBEL solution and it is nowadays one of the more popular boundary conditions since it is easily implemented into numerical schemes using iterative methods. Applications for plain journal bearings can be found e.g. in [21, 36, 153] and for porous journal bearings in [130, 133, 134, 139, 150].

Although the REYNOLDS condition has proved to be very effective for plain journal bearings, it does not assure mass balance between the fluid film and the cavitation area. The boundary condition proposed by JAKOBSSON, FLOBERG [73] and OLSSON [121] (JFO) corrects for this. The ELROD algorithm [45, 46] implements this JFO-conditions with a computational algorithm using a variable which represents both the fluid pressure in the fluid film and the fluid fraction in the cavitation area. Since the publication of the algorithm several applications to numerical schemes have been published, which also could be applied to plain journal bearings [86, 154, 160]. The Elrod algorithm was also applied to porous journal bearings by several authors [51, 103, 150]. Another mass-conserving approach is to consider the fluid film as a control volume and to determine the fluid film extent by using the momentum integral method [128], which was also applied to porous journal bearings [82, 83]. Recently the Elrod algorithm was adopted and modified for a so-called simulated-annealing technique which was applied to model porous journal bearings lubricated with ionic fluids [115].

For an extensive overview of cavitation in fluid film bearings the reader is referred to [19, 41].

Solution methods

To be able to predict the forces and moments in a plain journal bearing, the pressure in the fluid film has to be determined. As was mentioned before this pressure is governed by the Reynolds equation and solving this differential equation is therefore the central issue in this thesis. When dealing with a porous journal bearing additionally Darcy's law needs to be solved which now is coupled by an extra fluid exchange term in the Reynolds equation and pressure continuity between the fluid film and the porous bush.

---

[4]For special cases this is however a valid solution, for example in the so-called submerged journal bearing

For simplified cases for solid journal bearings – the so-called short and long bearing approximations [104, 119, 144] – the Reynolds equation can be solved analytically [25, 152]. When considering a porous bearing particular simple analytical solutions are available when the porous bush is assumed thin [23, 106, 130]. By using separation of variables the Darcy equation can be solved in Cartesian and polar coordinates [31, 102, 107, 108, 131, 138], which results in analytical solutions that are valid for porous journal bearings with a porous bush with finite thickness.

However, to solve the Reynolds equation for plain journal bearings of finite length one has to rely on either semi-analytical or numerical discretization methods. The Galerkin method with global ansatz functions is such an approach and has been applied by several authors to solve the Reynolds equation for solid journal bearings [52, 148]. By solving Darcy's law with separation of variables, this method can also be applied to porous journal bearings [31].

When the solution field is discretized Galerkin's method is also applicable as a numerical discretization method. Local ansatz functions instead of global functions are now used for each finite element to build a system of equations. This method is referred to as the finite element method (FEM) and can be applied to more complex geometries due the discretization. It has been applied to the plain journal bearing by several authors e.g. [36, 56, 86, 160] and is now a standard method available in commercial software [85]. Although some authors have used FEM for porous journal bearings [133, 150] the finite difference (FD) method seems to be preferred in literature. By approximating the differential operators in the Reynolds equation with finite differences a system of (linear) equations is obtained, which usually is solved using iterative algorithms such as Jacobi, Gauss-Seidel and Successive-Over-Relaxation (SOR). Due to their easy implementation and flexibility many authors have used FD methods to obtain the pressure in solid journal bearings e.g. [21, 152] and porous journal bearings [48, 50, 51, 77, 78, 81–83, 103, 134].

Porous-fluid interface

To capture the essential behavior in a porous journal bearing, it is crucial that flow between the porous bush and the fluid film is incorporated adequately. The most important interactions are continuity of pressure and flow in radial direction [23, 31, 106, 107, 131, 138]. Additionally, continuity of flow in axial and circumferential direction can be demanded at the interface [103, 134], however, for realistic parameters this influence proved to be negligible [103]. Physically more correct is to use the BEAVERS-JOSEPH boundary condition [9] to model the transition from the fluid film to the porous medium [57, 77, 79–83, 102, 108, 131]. However,

BEAVERS-JOSEPH's boundary condition results in a discontinuity in the velocity's derivative at the porous-fluid interface. BRINKMAN's model [20] assures this continuity and also was implemented for porous bearings [51, 93, 94, 96–100]. Nevertheless, even this sophisticated interface condition seems to have only a small impact on the bearing characteristics for realistic bearing parameters [55, 103]. Moreover is the choice of the parameter which governs the slip conditions problematic, since only experimental data are available for extremely simplified cases. An elaborate discussion on the choice of parameters and the validity of the different slip-flow boundary conditions is given in [117]. This topic will be discussed in more detail in section 2.2.

Rough surfaces

In reality the bearing surface will not be perfectly smooth, but have a certain roughness profile. This roughness will affect the hydrodynamical pressure and – more importantly – induce solid contact pressure due to elastic and plastic deformation of micro-asperities on the bearing surface.

It is possible to calculate the influence of these micro-asperities in real-time with direct pressure calculation of each contact, using the BOUSSINESQ-CERUTTI theory[5] [5, 75, 143]. A roughness profile is needed, which can be generated numerically or obtained from measurements. This approach has been applied in several studies on solid journal bearings e.g. [13, 156]. A more simplified approach is to describe the pressure induced by the asperity contacts statistically. GREENWOOD & WILLIAMSON [58, 59] proposed a model where the asperities are approximated by paraboloids, whose contacts can be conveniently described with Hertzian theory. Now only a few statistical parameters are needed to describe asperity contacts. The reader is referred to [10, 12] for an elaborate overview of multiple asperity contact properties and models.

Flow in the fluid film can also be obstructed by the micro-asperities, depending on their direction and shape. Using so-called flow factors these effects can be incorporated in the Reynolds equation. Calculation of these factors can be performed either analytically using stochastic models to describe the asperities [26, 27] or numerically considering every micro-contact in a small area [123–125]. More recently, homogenization techniques were utilized to calculate these flow factors to describe the surface roughness with more accuracy [2, 37, 159]. Thus far stochastic models have been used to model the influence of rough surfaces on the hydrodynamical pressure in porous journal bearings [65, 66, 68, 129].

---

[5]Often this theory is also referred to as the model of elastic half-space or simply as the BOUSSINESQ theory

Elastohydrodynamic lubrication

When high pressures are generated in a solid journal bearing its bush may notably deform. This will in turn affect the film thickness function and therefore influence the characteristics of the fluid film. Because an additional differential equation is introduced, the complexity of the problem increases significantly.

Assuming the deformation to be small and purely elastic, the so-called thin liner model can be used. This model assumes that elastic deformation only takes place in a thin liner in the inner edge of the bush [69, 70, 152]. When larger deformations occur these are usually calculated using FEM [13, 84, 120].

Although the generated pressure and the applied load for porous bearings are considerable lower in comparison to solid journal bearings, some studies on flexible porous journal bearings are available [50, 95, 101].

Misalignment

Several authors studied the influence of misalignment in solid journal bearings e.g. [14, 74, 127, 155] and its effect on stability [118]. To be able to accurately predict the fluid film pressure for misaligned journal bearings with very small minimum film thickness many degrees of freedom are needed. It has however been shown that for normal operating conditions simpler modeling approaches with less degrees of freedom seem to be adequate [15].

Few studies address misalignment in porous journal bearings and these only account for steady state situations using numerical discretization methods [47, 133].

Non-Newtonian fluids

Because most porous bearing applications deal with light loading and pressures, the use of non-Newtonian fluids and their dependency on density, pressure and temperature has not been broadly studied. Several models to describe non-Newtonian fluids are available, an overview is given in e.g. [13]. Some studies account for the effect of percolation of polar additives into the porous matrix, which happens when so-called couple stress fluids are used for lubrication in porous journal bearings [49, 92, 111–114]. Another improvement is to include the viscosity dependency on pressure using BARUS' law [51].

Experimental evidence

Cameron et al. [22] were the first to compare experimental with theoretical results. In their work the theoretical results of the short bearing model from [106] were corrected to finite length and friction values were compared. Although the results are in good agreement for lower loads, a large deviation is seen when the load exceeds a critical value. The authors ascribe this discrepancy to the effect of pore closure which occurs at high eccentricity ratios and is not accounted for in their model.

Using optical interferometry Yung et al. [161] studied the film formation and running-in behavior at the bearing surface in detail. Small so-called micro-land pads are observed in the load carrying zone in between which ruptures the fluid film. An essential conclusion was that the running-in behavior allows for sufficiently flat surfaces that are necessary for film formation and therewith hydrodynamic lubrication.

Mokhtar et al. [105] show comparable results for friction coefficients, but also compared experimental and theoretical journal center loci from [106] for increasing loads. Here the trend in change of attitude angle is confirmed when permeability is varied, however, a significant discrepancy still remains.

The oil film extent of a porous bronze bearing was studied using a fluorescent-dyed oil combined with ultraviolet lamps and a shaft made of transparent silica glass in [81]. This showed that there is significant difference in oil film extent between solid and porous journal bearings. It is concluded that the angular oil film extent is considerably smaller than for solid journal bearings. Subsequent papers of the same authors build on this conclusion and a corresponding model was developed [78, 80, 83]. Although good agreement of this model with experiments was shown, an important addition to all experimental setups was a circumferential groove in the porous bush with an oil feed pressure applied to it. It seems that the oil pressure is an important, influential model parameter. However, most normal porous journal bearings do not have an external oil supply or circumferential groove.

More recently Tschentke [150] studied the porous journal bearing by comparing experimental results to simulation results using an FEM based model which utilizes JFO cavitation proposed in [86]. To some extent the model agrees to the experimental results, but it is noted that these results depend strongly on several model parameters. These include the permeability, its distribution throughout the porous matrix and the cavitation pressure. The permeability distribution is influenced during the fabrication process and it also changes over time due to running-in. The cavitation pressure seems to be especially problematic, because an unrealistic value is needed to achieve a good comparison. Similar behavior has been observed in [103].

Some studies have been published that investigate the wear mechanisms in porous journal bearings experimentally [43, 44]. The influence of periodic loading on the friction coefficient of porous journal bearings was studied in [132].

Nonlinear dynamics of rotor-bearing systems

In reality, a rotor-bearing system will possess a certain unbalance and usually carries a dynamical load (often with changing direction). When this is the case the journal center's motion will – depending upon the loading situation – either converge with a whirling motion to a steady state solution or become unstable. One way to quantify the dynamic behavior is to linearize the forces generated by the pressure, which results in linear stiffness and damping coefficients. These coefficients only hold for a specific working point e.g. defined by eccentricity and are only valid for small changes in position and velocity around this point. Using stiffness and damping coefficients one can identify the linear stability threshold of the steady state [54]. This gives a first insight at which point the journal motion becomes unstable and when critical rotor speeds are reached. Many references describe the procedure in obtaining the stiffness and damping coefficients of plain journal bearings, e.g. [54, 152]. Some studies with analytical expressions for stiffness and damping coefficients in porous journal bearings for the simplified short and long bearing cases are available [24, 29, 32, 33, 63, 98]. In other studies the problem is solved numerically and the coefficients are calculated for the general (finite bearing) case [62, 63, 77, 110].

For large deflections linearizations are no longer applicable and one has to use the nonlinear force relationship. This regime now has to be studied using numerical time integration, steady-state solvers to obtain periodical solutions and path followers. A detailed survey is beyond the scope of this thesis and the reader is referred to e.g. [152] for an overview. In this thesis the software MATCONT [39] has been used for path-following and the analysis of bifurcations, however, other software such as AUTO [40] or BIFPACK [141] can also be used here.

Using these methods it is possible to identify fixed-point and periodic solutions and their dependence on system parameters such as angular frequency, unbalance or – in the case of a porous journal bearing – permeability. Therewith so-called bifurcation diagrams can be created, that represent the dynamical solution of the nonlinear system in a compact way. These techniques can be applied to rotor-bearing systems containing plain journal bearings e.g. [30, 71, 152]. Only few publications are available that deal with the nonlinear dynamical behavior in porous journal bearings [89, 90, 99].

1.3 Thesis outline

The first part of this thesis introduces the underlying governing equations describing the physical phenomena that are relevant for the plain journal bearing problem. This includes the theory of lubrication in thin films and flow through porous media and are respectively governed by Reynolds' equation and Darcy's law. Several corresponding effects such as cavitation conditions in the fluid film and slip-velocity conditions at the porous-fluid interface are discussed as well as the influence of surface roughness on the fluid pressure and asperity contacts.

The following chapter focuses on dynamical systems containing plain journal bearings and includes the subjects of multibody systems and nonlinear dynamics. The kinematics and dynamics of revolute joints with clearance (i.e. plain journal bearings) will be discussed, which is followed by a brief introduction of steady state solutions, (non)linear stability and potential bifurcation points that can occur in the dynamical systems that are studied in this work.

In the second part the reader is introduced to the plain journal bearing and several modeling approaches to calculate the forces and moments generated in journal bearings. First the geometry, the governing equations and the boundary conditions of the plain journal bearing are given as well as a short explanation of the impedance method and definitions of the forces and moments. After this analytical, semi-analytical and numerical discretization methods for plain journal bearings are discussed. Here, the Galerkin method is proposed as a semi-analytical approach to solve the dynamical solid and porous journal bearing problem. Along with this a description to calculate stiffness and damping coefficients and an approach to include the influence of surface roughness with flow factors are given.

The following chapter briefly focuses on the performance and accuracy of the proposed bearing model. After that the Galerkin-based bearing model will be verified through comparison with existing analytical solutions for the solid journal bearing and by comparison with numerical methods for solid and porous journal bearings. Finally, friction curves of porous journal bearings are compared to experimental data to establish the validity of the proposed bearing model. Continuing, this chapter will then focus on the characteristics of a single plain journal bearing. The behavior of the porous journal bearing with respect to a solid journal bearing will be shown and typical characteristics of plain journal bearings such as damping and stiffness coefficients and impedance maps are studied. These also include the influence of surface roughness on the hydrodynamic pressure and asperity contacts. Finally, the consequences of misalignment for the force and moment capacity are shown.

In the last part rotor-bearing systems containing plain journal bearings are studied. The onset of instability as well as the occurrence of bifurcation points is investigated for elementary (un)balanced rotor-bearing systems containing porous and solid journal bearings and a rigid rotor. The influence of misalignment is discussed for a rotor-bearing system containing a flexible rotor. Finally, the applicability of the developed bearing model for real live engineering problems is demonstrated with a realistic multibody system simulated with the multibody software ADAMS.

The final chapter contains a conclusion and an outlook, including the most important results and recommendations for future research.

Chapter 2

Lubrication theory

This chapter describes the theory of conformal lubricated contacts appearing for example in the plain journal bearing which will be introduced in detail in chapter 4. The laws governing flow in thin films and porous media are introduced as well as relations describing the influence of surface roughness on the (dynamical) behavior of a lubricated contact.

2.1 Fluid film lubrication

The contact occurring in a plain journal bearing is a conformal contact, which means that the contact surface curvatures are very similar to one another. Figure 2.1 shows such a lubricated contact in a Cartesian coordinate system. The arbitrary velocities U_1, V_1, W_1 and U_2, V_2, W_2 are prescribed at the fluid boundaries at the bottom and top surfaces respectively. These velocities are imposed to the fluid by the motion of the confining surfaces and may also include additional components when the fluid can also flow through the surface. The distance between the surfaces is given by the film thickness function $H(x,z)$. When making the assumptions that

- the lubricant matches the prescribed velocities at both contact surfaces (which can include additional components when the fluid is allowed to also flow through the surface),

- the fluid behaves like a Newtonian fluid with constant density ρ and has a constant dynamical viscosity η,

- the film is sufficiently thin (i.e. curvature and temperature differences can be neglected),

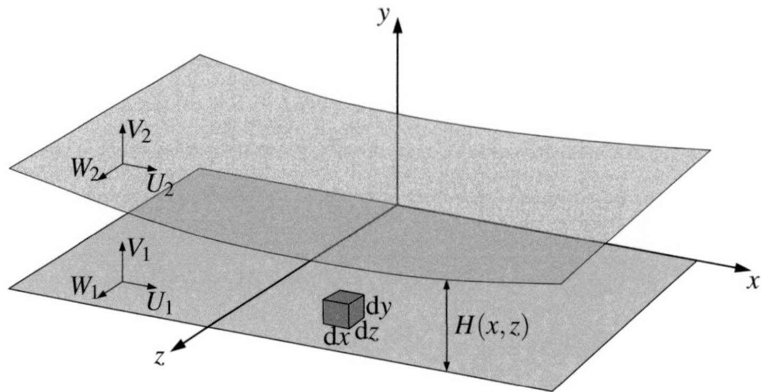

Figure 2.1: Lubrication of a conformal contact with a thin fluid film having film thickness $H(x,z)$.

- in the region where the fluid film forms, it extends continuously from the top to the bottom surfaces,

- inertial forces and turbulent behavior in the fluid can be neglected (i.e. the fluid flow is laminar)

the Reynolds equation can be deduced from the Navier-Stokes equations to describe the hydrodynamic gage pressure p_h in the fluid film[1]. The Reynolds equation reads [147]

$$\frac{\partial}{\partial x}\left(\frac{H^3}{12\eta}\frac{\partial p_\mathrm{h}}{\partial x}\right)+\frac{\partial}{\partial z}\left(\frac{H^3}{12\eta}\frac{\partial p_\mathrm{h}}{\partial z}\right)=H\frac{\partial}{\partial x}\frac{U_1+U_2}{2}+\frac{U_1-U_2}{2}\frac{\partial H}{\partial x}\cdots$$
$$\cdots+H\frac{\partial}{\partial z}\frac{W_1+W_2}{2}+\frac{W_1-W_2}{2}\frac{\partial H}{\partial z}+(V_2-V_1).$$

$$(2.1)$$

Please refer to figure 2.1 for the other used parameters and variables. The reader is referred to appendix A for a detailed derivation based on the force balance over an infinitesimal volume element.

The main issue in this thesis will be to solve this partial differential equation in order to obtain the hydrodynamical pressure in the fluid film, from which the majority of important quantities can be deduced. To be able to do this proper boundary conditions need to be applied to the pressure field. Aside from the geometrical boundary conditions which bound the pressure function in the fluid film, the most important other condition occurs due to cavitation.

In order to apply the Reynolds equation to journal bearing problems the velocities U_1, V_1, W_1

---

[1]Unless stated otherwise all references to fluid pressure in this thesis refer to the so-called gage pressure, which is the absolute pressure minus atmospheric pressure.

and U_2, V_2, W_2 need to properly represent the motion of the journal and the bearing in the same reference frame, which will be shown in chapter 3. The geometrical boundary conditions are related to the journal bearing type and will be introduced in chapter 4, the subject of cavitation will be discussed in the following.

2.1.1 Cavitation

Figure 2.1 illustrates a typical lubricated contact in which cavitation can occur. Imagine a fluid flowing in the positive x-direction. Due to fluid and pressure build-up in the converging part of the fluid film pressure in the diverging part will drop significantly. If the pressure does not fall below the vapor pressure, e.g. because the surrounding pressure is raised using an external fluid supply, an additional cavitation condition is not needed and it will suffice to solve for the pressure p_h from equation (2.1) solely using boundary conditions that geometrically bound the pressure field. In general, however, this is not the case and a serious error is made when the negative pressure part is included in succeeding calculations. Under normal (hydrodynamical) operating conditions film rupture occurs in plain journal bearings somewhere downstream of the minimum film thickness. Instead of fluid there now will be a mixture of gas and fluid (or so-called two-phase flow) throughout a certain extent of the clearance until a certain point where the fluid film will reform, as shown in [41, 42]. This usually is where the film thickness function starts to converge again. The pressure throughout this cavitation area is negligible and it therefore strongly influences the load capacity of the journal bearing. This section will present the most common cavitation models from literature, which can be categorized in two categories: non-mass-conserving and mass-conserving boundary conditions.

Non-mass-conserving cavitation conditions

Three typical pressure profiles for a lubricated conformal contact can be found in figure 2.2. This figure shows the pressure profile obtained for a steady state case[2] with different cavitation conditions. Without any additional condition the SOMMERFELD solution is obtained. As already has been mentioned the negative part of this solution is unrealistic and should in general not be taken into account for the succeeding calculations.

There are several improved boundary conditions to correct for this. The simplest of them is the

---

[2]By steady state is meant that the time dependency of the film thickness function is negligible i.e. V_1 and V_2 in equation (2.1) can be ignored. This does, however, not mean that $U_{1,2}$ and $W_{1,2}$ necessarily are zero!

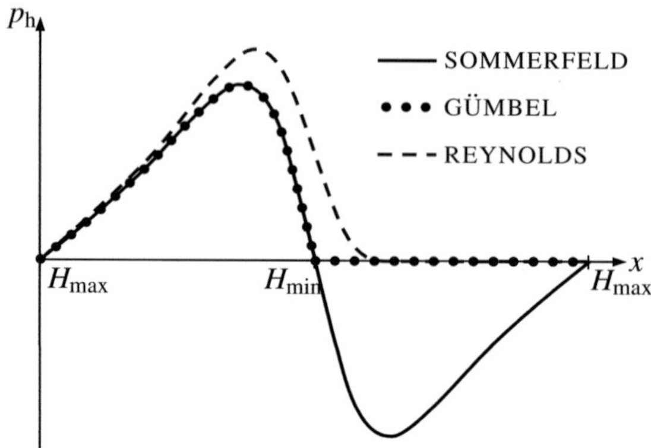

Figure 2.2: Theoretical steady state pressure profiles in a lubricated conformal contact for different boundary conditions. The cavitation pressure is here set to zero ($p_\mathrm{cav} = 0$).

so-called HALF-SOMMERFELD or GÜMBEL solution [64]. This method only takes into account the positive part of the SOMMERFELD solution, i.e.

$$p_\mathrm{h}(x > H_\mathrm{min}) = 0. \tag{2.2}$$

Due to its simplicity this condition is still applied in many theoretical studies, including this work. It should be mentioned that in this form this cavitation condition is not entirely correct when addressing dynamic problems. This issue will be discussed later on.

For steady state conditions the REYNOLDS or SWIFT-STIEBER boundary condition [145, 146] is believed to be more accurate. Here the pressure gradient is demanded to vanish where the fluid film ruptures. This condition is implemented with the additional boundary conditions

$$\left.\frac{\partial p_\mathrm{h}}{\partial x}\right|_{p_\mathrm{h}=p_\mathrm{cav}} = \left.\frac{\partial p_\mathrm{h}}{\partial z}\right|_{p_\mathrm{h}=p_\mathrm{cav}} = 0, \tag{2.3}$$

where p_cav indicates the cavitation pressure. This solution is a good approximation of reality, especially at larger eccentricities [147]. Moreover is the pressure function's derivative continuous at the rupture boundary, which indicates that mass is being conserved here. This is, however, still not the case for the reformation point (at H_max), which will be addressed in the following paragraph. When the GÜMBEL condition is compared to the REYNOLDS condition a difference is observed. The solution with the GÜMBEL condition seems, however, to be a reasonable approximation. We shall see in chapter 5 that the difference actually turns out to be quite small for typical plain journal bearings geometries. Unfortunately, the REYNOLDS condition is harder to implement for analytical methods. The problem complicates to a free boundary problem in which the rupture and reformation points of the fluid film are a priori unknown. Implementation into a numerical iteration scheme is, on the other hand, easier. Such algorithms will be discussed in chapter 4.

Mass-conserving cavitation conditions

Although the REYNOLDS condition is mass-conserving at the rupture boundary, it still is not completely mass-conserving over the entire circumference of the bearing. At the fluid film reformation point the amount of fluid, which comes from the cavitation range is not taken into account. The first mass-conserving cavitation conditions were introduced by JAKOBSSON, FLOBERG [73] and later by OLSSON [121] for dynamical problems. These cavitation conditions therefore are conveniently abbreviated as the JFO conditions. Floberg proposed boundary conditions for the vapor or cavitation region. Later on Elrod [45] introduced an algorithm which utilizes a modified Reynolds equation to automatically distinguish between the fluid film and the cavitation range. Due to its complexity and high computational demands this cavitation algorithm has not been employed in this work and because of this a detailed description is omitted here.

Dynamic problems

Up to this point, all the introduced cavitation conditions only were formulated for steady state problems, for which the velocity term $V_2 - V_1$ on the right hand side in equation (2.1) could be ignored.

Figure 2.3 shows that when the change of the film thickness function over time becomes more profound, the boundaries of the positive pressure range of the GÜMBEL solution do not necessarily coincide with the minimum (H_{min}) and maximum film thickness (H_{max}) (as was the case in figure 2.2). As a result the boundaries of the positive pressure range have to be determined additionally. For some simplified analytical plain journal bearing models analytical expressions for these boundaries can be formulated, as will be shown in chapter 4.

The REYNOLDS condition for the rupture boundary still holds for dynamical problems. However, problems arise for the fluid film formation boundary, because its position is not known a priori anymore. One solution is to also apply the REYNOLDS conditions to the formation boundary. When doing so the pressure derivative will also be continuous at this boundary. The JFO condition shows, however, that in general a discontinuity does occur here and applying the REYNOLDS condition to both boundaries might be unrealistic [158].

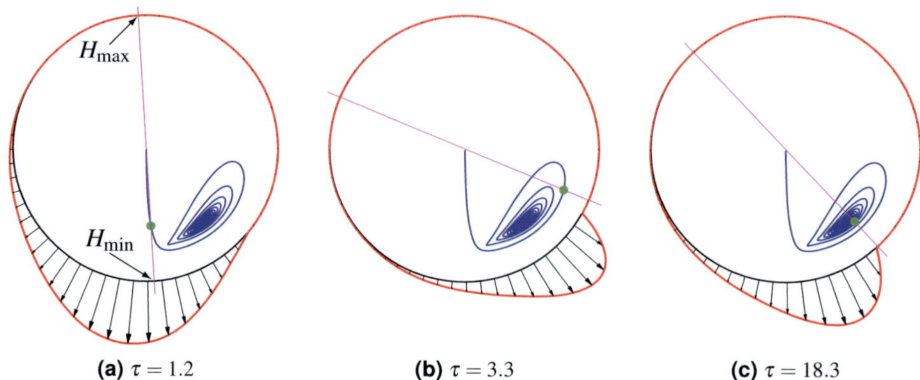

(a) $\tau = 1.2$ **(b)** $\tau = 3.3$ **(c)** $\tau = 18.3$

Figure 2.3: Snapshots of the dynamical pressure profile of a so-called short bearing model (the reader is referred to section 4.2 for a detailed description of this model). The journal starts in the center and moves to an equilibrium position due to loading in vertical direction. The red curves corresponds to the pressure, the blue curve shows the orbit, the green dot the journal center's position and the maximum (H_{max}) and minimum (H_{min}) film thickness are indicated with the magenta line of centers.

Although the JFO boundary condition is more accurate, the corresponding methods require too much calculation time to be applicable to the problems in this work. Therefore, this work will mainly employ the GÜMBEL condition to solve the plain journal bearing problem.

2.1.2 Mixed lubrication

The phenomenon of mixed lubrication is the combined influence of hydrodynamic lubrication, its dependency on rough surfaces and the influence of asperity contacts that occur due to surface roughness. In the previous section all contact surfaces were assumed to be completely smooth, which is a valid assumption for large fluid film thicknesses compared to the asperity peaks heights. For small film thicknesses this influence has to be taken into account: within this thesis the flow factor method proposed by PATIR and CHENG [123–125] to include the influence of rough surfaces on the hydrodynamic pressure is applied. The asperity contacts itself are modeled with the statistical contact model proposed by GREENWOOD & WILLIAMSON [58, 59]. To be able to describe the previously mentioned phenomena knowledge about the statistical properties of the roughness profile is required. The rough surface is defined by adding a deviation δ_i (where the subscript i indicates the surface) to each nominal contact surface as is illustrated in figure 2.4. This leads to a new local film thickness function, defined by H_T;

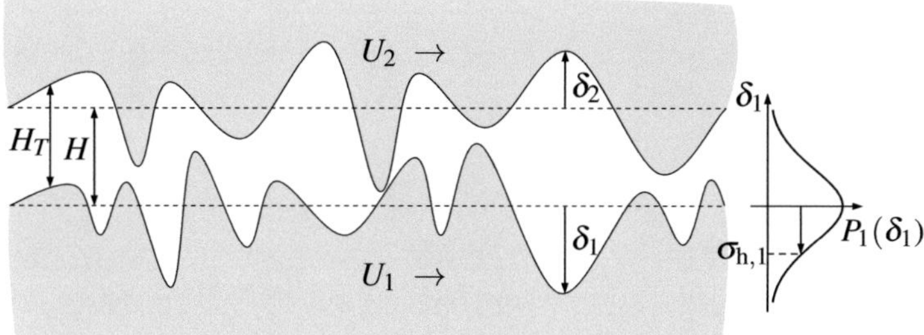

Figure 2.4: Definition of a contact between rough surfaces.

$$H_T(x,z) = H(x,z) + \delta_1(x,z) + \delta_2(x,z). \tag{2.4}$$

In a statistical sense, the variation of the roughness profiles can be characterized with a corresponding root mean square

$$\sigma_{h,i}^2 = \frac{1}{A}\int_A \delta_i(x,z)^2 \mathrm{d}A, \tag{2.5}$$

where A corresponds to the nominal contact area. This also is the standard deviation of the probability density function $P_i(\delta_i)$ if δ_i corresponds to the deviation from the mean surface distance. From this the probability function can be defined with

$$F(d) = \int_d^\infty P(\delta)\mathrm{d}\delta, \tag{2.6}$$

which gives the percentage of asperities that are in contact for a mean surface distance d. When treating problems where both surface roughnesses play a role, the problem is replaced by an equivalent problem with a composite standard deviation $\sigma_h = \sqrt{\sigma_{h,1}^2 + \sigma_{h,2}^2}$. Apart from these statistical parameters describing the vertical surface heights another important characteristic of the surface are its directional properties. Depending on the flow direction a so-called Peklenik factor $\tilde{\gamma}$ can be defined, using the characteristic length scales $\lambda_{x,z}$ (in x- and z-direction) to which the auto-correlation function of the roughness profile reduces with a certain percentage of its initial value [126]. The auto-correlation function is defined by

$$R_{\mathrm{cor}}(\lambda_x,\lambda_z) = \frac{1}{\sigma_h^2}\lim_{A\to\infty}\frac{1}{A}\int_A \delta(x,z)\delta(x+\lambda_x,z+\lambda_z)\mathrm{d}A \tag{2.7}$$

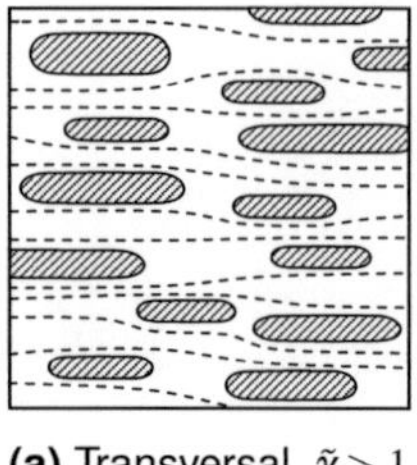

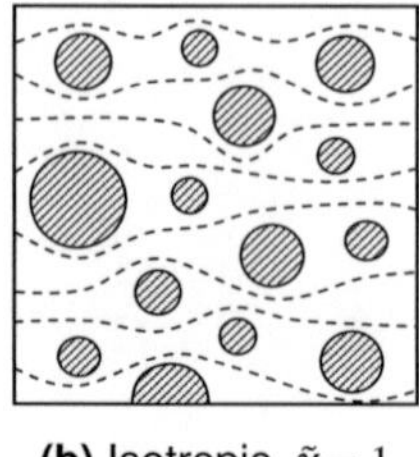

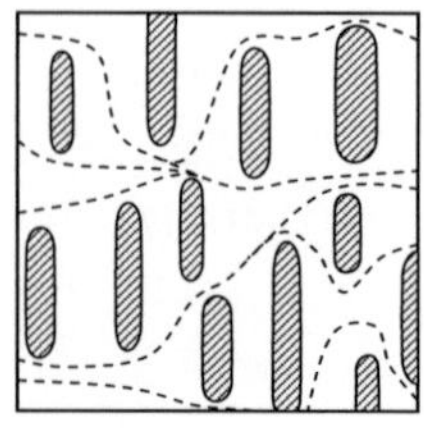

(a) Transversal, $\tilde{\gamma} > 1$ **(b)** Isotropic, $\tilde{\gamma} = 1$ **(c)** Longitudinal, $\tilde{\gamma} < 1$

Figure 2.5: Contact areas for three roughness profiles with different Peklenik factors for a given flow direction. Here, the x-axis points in horizontal direction and the z-axis in vertical direction.

and with this the Peklenik factor is defined by

$$\tilde{\gamma} = \frac{\lambda_x(R_{\mathrm{cor}} = 0.5)}{\lambda_z(R_{\mathrm{cor}} = 0.5)}. \tag{2.8}$$

Here, the first index of the characteristic length scales indicates the amount to which the auto-correlation function is reduced, in this case a percentage of 50%. Depending on the roughness profile different values can or have to be chosen here. Figure 2.5 illustrates the influence of the Peklenik factor on the contact area with a given flow direction.

Hydrodynamic lubrication of rough surfaces

Although possible, it is hard to include the original surface roughness profile into the numerical simulation (it would require extremely fine grids, micro-cavitation algorithms, etc.). As will be shown in chapter 4, numerical discretization methods alone already demand too many degrees of freedom. Therefore, Patir et al. [123–125] proposed an averaged flow model to account for the surface's roughness. The idea is to take a small patch of the rough surface which then is taken to represent the entire rough surface. Using a small scale hydrodynamic simulation an averaging procedure is introduced, which produces so-called flow factors. The flow factors account for the fact that the asperities have modified the effective film thickness function. In turn, these flow factors can then be used to construct an averaged Reynolds equation which incorporates an approximation of the surface roughness influence:

$$\frac{\partial}{\partial x}\left(\phi_x^f \frac{\bar{H}^3}{12\eta} \frac{\partial \bar{p}_{\mathrm{h}}}{\partial x}\right) + \frac{\partial}{\partial z}\left(\phi_z^f \frac{\bar{H}^3}{12\eta} \frac{\partial \bar{p}_{\mathrm{h}}}{\partial z}\right) = \frac{U_1 + U_2}{2} \frac{\partial(\bar{H})}{\partial x} + \frac{U_1 - U_2}{2} \sigma_{\mathrm{h}} \frac{\partial \phi_s^f}{\partial x} + (V_2 - V_1),$$
$$\tag{2.9}$$

where ϕ_x^f and ϕ_z^f are so-called pressure flow factors and ϕ_s^f is the shear flow factor. $\bar{H}$ and $\bar{p}_h$ respectively are the mean film thickness and mean fluid pressure. Note that for reasons of simplicity the velocity terms in the z-direction have been omitted here. To determine the pressure flow factors a measured or statistically generated roughness profile is used for a flow simulation on a small yet representational surface patch (i.e. including a sufficient number of asperities). Now say that the macroscopic film thickness in the contact will equal $\bar{H}$ and be constant and let the microscopic film thickness function be H_{T}. Then the pressure flow factors are determined using

$$\phi_x^f = \frac{\dfrac{1}{L_z}\displaystyle\int_0^{L_z}\left(\dfrac{H_{\mathrm{T}}^3}{12\eta}\dfrac{\partial p_h}{\partial x}\right)\mathrm{d}z}{\dfrac{\bar{H}^3}{12\eta}\dfrac{\partial \bar{p}_h}{\partial x}} \quad\text{and}\quad \phi_z^f = \frac{\dfrac{1}{L_x}\displaystyle\int_0^{L_x}\left(\dfrac{H_{\mathrm{T}}^3}{12\eta}\dfrac{\partial p_h}{\partial z}\right)\mathrm{d}x}{\dfrac{\bar{H}^3}{12\eta}\dfrac{\partial \bar{p}_h}{\partial z}}, \tag{2.10}$$

where $L_{x,z}$ indicate the geometrical sizes of the surface patch and $\bar{p}$ the average macroscopic pressure. The shear flow factor is determined by

$$\phi_s^f = \frac{2}{U_s\sigma_h}\bar{q}_x = \frac{2}{U_s\sigma_h}\mathcal{E}\left(-\frac{H_{\mathrm{T}}^3}{12\eta}\frac{\partial p_h}{\partial x}\right), \tag{2.11}$$

where $U_s = U_1 - U_2$ and $\mathcal{E}(\dots)$ the expectation operator. The expectation operator is for instance used to determine the mean film thickness; $\bar{H} = \mathcal{E}(H_{\mathrm{T}})$.

Since this thesis is only concerned with the modeling of plain journal bearings in the context of larger dynamical systems, the calculation of flow factors is beyond the scope of this work. In chapter 4 a general implementation for flow factors is proposed and the examples in chapter 5 feature the flow factors published in [124, 125] as a prove of concept.

Asperity contact

GREENWOOD and WILLIAMSON introduced an asperity contact model which is based on statistical parameters, it was later extended for two rough surfaces coming into contact [58, 59]. The idea behind this model is to simplify the roughness profile, approximating it with a surface that is solely made up out of identical parabolic asperities, possessing a curvature with radius $\hat{\beta}$, the standard deviation σ_e of the peak heights and the surface density $\hat{\eta}$ of the asperity peaks on the surface. Using the Hertzian approximation for the contact of one asperity peak the entire surface roughness can be approximated. The resulting load and contact area from a single elementary asperity can now be described by

$$A_\mathrm{e} = \pi \frac{\hat{\beta}}{2} w \quad \text{and} \quad P_\mathrm{e,load} = \frac{4}{3} E \sqrt{\hat{\beta}/2} w^{3/2}, \tag{2.12}$$

where w is the indentation depth of the asperity and E is the composite elastic modulus[3] of both contact surfaces (the subscript e refers to the elastic contact). Using this load function the nominal pressure (i.e. the complete load $P_\mathrm{e,area}$, per nominal contact area A_nom which can be described with the previous introduced roughness parameters) for a rough to smooth surface contact can be expressed with

$$p_\mathrm{e}(H) = \frac{2\sqrt{2}}{3}(\hat{\eta}\hat{\beta}\sigma_\mathrm{e})E\sqrt{\frac{\sigma_\mathrm{e}}{\hat{\beta}}} F_{3/2}\left(\frac{H}{\sigma_\mathrm{e}}\right) \tag{2.13}$$

and for a rough to rough surface contact with

$$p_\mathrm{e}(H) = \frac{16\sqrt{2}}{15}\pi(\hat{\eta}\hat{\beta}\sigma_\mathrm{e})^2 E\sqrt{\frac{\sigma_\mathrm{e}}{\hat{\beta}}} F_{5/2}\left(\frac{H}{\sigma_\mathrm{e}}\right), \tag{2.14}$$

where $F_n(d)$ is the probability density function of the asperity peaks (analogue to equation (2.6)),

$$F_n(d) = \int_d^\infty (u-d)^n s(u)\,\mathrm{d}u, \tag{2.15}$$

in which $s(u) = 1/\sqrt{2\pi}\exp\left(-u^2/2\right)$ for a Gaussian height distribution. By using the error function, this integral can be approximated with a simplified expression for Gaussian distributions. The approximation reads

$$F_n(d) \approx c_0\left(1 - \mathrm{erf}(\lambda_1 d + \lambda_2)\right), \tag{2.16}$$

where $c_0 = 3.851$, $\lambda_1 = 0.478$ and $\lambda_2 = 1.123$ for the rough to smooth surface contact ($n = 3/2$) and $c_0 = 23.375$, $\lambda_1 = 0.454$ and $\lambda_2 = 1.563$ for the rough to rough surface contact ($n = 5/2$). This approximation has the advantage that it is valid for arbitrary film thicknesses. Other approximations are described in [88, 123].

The original GREENWOOD & WILLIAMSON approach gives good results when the probability density function of the asperity peaks is Gaussian. However, due to prolonged loading running-in can occur, which can change the shape of the probability density function significantly. For such cases either a more accurate distribution function or a more accurate method to calculate

---

[3]The composite elastic modulus E is determined by $1/E = (1-v_1^2)/E_1 + (1-v_2^2)/E_2$, where $E_{1,2}$, $v_{1,2}$ denote resp. the elastic modulus and Poisson's ratio of the contact bodies.

the local contact pressure is required. Since determining the distribution change after prolonged exposure can be difficult and unpractical, the usual approach is to use the BOUSSINESQ theory to calculate the pressure and deformations resulting from asperity contacts numerically [10]. Using this approach it is possible to simulate the elastic asperity deformations and extend it to account for perfect plastic deformations [13]. Although it would be possible to numerically calculate the elastic contact behavior of a ran-in surface prior to a dynamical simulation, doing so for plastic deformations is very hard. In comparison, the GREENWOOD & WILLIAMSON approach utilizes only several statistical contact parameters (but only accounts for elastic deformations). Due to this simplicity the latter contact model is used in this work.

2.2 Lubrication of porous media

Next to the pressure in the fluid film, the fluid pressure in the porous matrix is also required when considering a porous journal bearing. To this end Darcy's law for flow inside a porous medium is introduced and the issues related to fluid-porous interfaces are addressed.

2.2.1 Darcy's law

By means of empirical methods, DARCY [34] concluded from his experiments on the flow of water in vertical homogeneous sand filters, that the rate of flow is proportional to the constant cross-sectional area and to the piezometric head[4] and that it is inversely proportional to the length of the porous medium. When combined these observations resulted in Darcy's law. The original formula was stated for one dimensional flow with an isotropic porous medium. For three dimensional flow in an anisotropic medium the relationship reads

$$\widehat{\boldsymbol{q}} = -\mathscr{H} \operatorname{grad} \boldsymbol{\varphi}, \tag{2.17}$$

where bold symbols indicate vectors and all quantities are macroscopic, $\widehat{\boldsymbol{q}}$ refers to the fluid velocity and grad $\boldsymbol{\varphi}$ represents the hydraulic gradient. This is a gradient of the piezometric head over the length of the flow path. $\mathscr{H}$ is the hydraulic conductivity tensor that expresses the ease with which the fluid is transported through the porous matrix. It depends on fluid properties such as the density ρ and viscosity η and properties of the porous matrix (introduced as the permeability $\mathscr{K}$ in the following) [8, 72]. The current formulation is practical when treating for

---

[4]The piezometric head is the sum of the pressure head and the elevation head, which are the pressure energy and the potential energy per unit weight of the fluid respectively.

example aquifers or ground water reservoirs, but less insightful when applied to porous journal bearings. A more useful formulation is

$$\widehat{\boldsymbol{q}} = -\frac{\mathscr{K}}{\eta}\left(\operatorname{grad}\widehat{\boldsymbol{p}} - \rho\boldsymbol{g}\right) \tag{2.18}$$

where the term $(\rho\boldsymbol{g})$ on the right hand side occurs due to gravity. More interesting though is $\mathscr{K}$, which represents the permeability tensor. The permeability defines the ability of the porous medium to transmit fluid. Throughout this thesis, the permeability will be assumed isotropic and homogeneous, for which the permeability tensor $\mathscr{K}$ reduces to a single scalar K [m^2]. To gain more insight in the physical meaning of the permeability it can be rewritten to $K = \tilde{N}\tilde{d}^2$, where $\tilde{d}$ [m] is a mean conduit diameter that characterizes the size scale of the pore structure and $\tilde{N}$ [-] is a shape factor which relates to the shape of the passages or conduits through which the fluid flows. Thus, increasing the porosity of a permeable material will also increase the mean conduit diameter and therefore also its permeability [72]. There are several ways to derive Darcy's law including complex mathematical derivations involving homogenization methods, for more details the reader is referred to e.g. [8, 72].

It should be noted that Darcy's law does not account for inertia effects as well as viscous friction within the fluid. To correct for this several extensions can be found in literature. For example

$$\widehat{\boldsymbol{q}} = -\frac{\mathscr{K}}{\eta}\left(\operatorname{grad}\widehat{\boldsymbol{p}} - \rho\boldsymbol{g}\right) + \rho\mathscr{C}\frac{\partial\widehat{\boldsymbol{q}}}{\partial t}, \tag{2.19}$$

where $\mathscr{C}$ is a so-called acceleration coefficient tensor. In most cases the time derivative can however be ignored and equation (2.19) reduces to the original form [117]. Another issue with Darcy's law is that it only holds for small fluid flow velocities $\widehat{\boldsymbol{q}}$. This can be corrected by introducing the so-called Forchheimer correction

$$\widehat{\boldsymbol{q}} = -\frac{\mathscr{K}}{\eta}\left(\operatorname{grad}\widehat{\boldsymbol{p}} - \rho\boldsymbol{g}\right) - F(\widehat{\boldsymbol{q}}), \tag{2.20}$$

where the last term on the right depends on the flow velocity $\widehat{\boldsymbol{q}}$ and can be defined in several ways depending on fluid and material properties. When the fluid velocity remains low, i.e. the Reynolds number Re does not exceed 10, Darcy's law still holds [8, 117].

Finally, the basic continuity for flow inside a porous medium is given for the general case of an inhomogeneous anisotropic porous medium in a volumetric average sense. Assuming the axes of the Cartesian coordinate system are aligned with the principal directions of the permeability

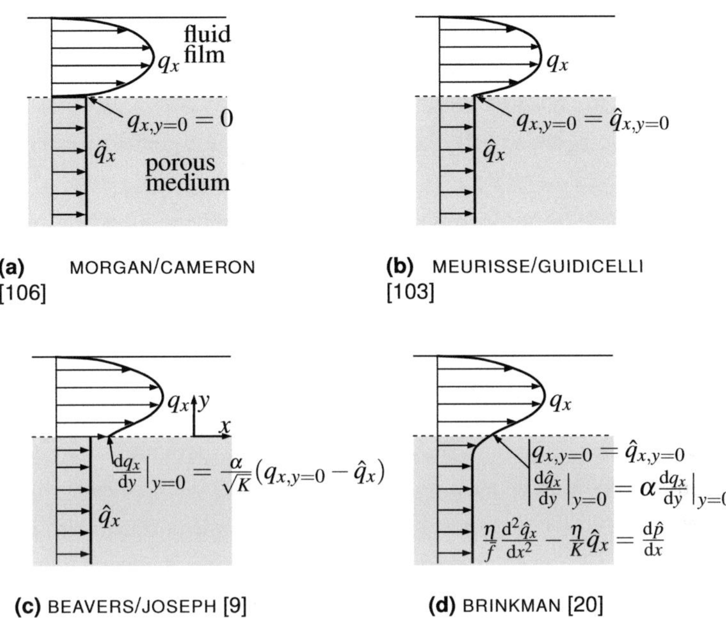

Figure 2.6: Different flow boundary conditions and the resulting fluid velocities $q_{x,y}$ (fluid film) and $\hat{q}_{x,y}$ (porous medium) near the interface for uniform flow through a channel partly filled with porous material [55].

tensor the following equation is obtained

$$\frac{\partial}{\partial x}\left(\frac{K_x \rho}{\eta}\frac{\partial \hat{p}}{\partial x}\right) + \frac{\partial}{\partial y}\left(\frac{K_y \rho}{\eta}\frac{\partial \hat{p}}{\partial y}\right) + \frac{\partial}{\partial z}\left(\frac{K_z \rho}{\eta}\left[\frac{\partial \hat{p}}{\partial z} + \rho g\right]\right) = \bar{f}\frac{\partial \rho}{\partial t}, \qquad (2.21)$$

where $\bar{f}$ indicates the porosity.

2.2.2 Porous-fluid interface

When a fluid flows through and alongside a porous medium, the fluid velocity will gradually change inside a transition zone at the porous-fluid interface. This transition is caused by viscous shear stresses in the fluid, which were deemed insignificant compared to the resistance of the porous medium in Darcy's law. Customizations are necessary to reintroduce this gradual transition into the problem formulation.

Several boundary conditions for the porous-fluid interface found in porous journal bearings were proposed in literature (see figure 2.6) [9, 20, 103, 106]. For reasons of simplicity the porous-fluid interface problem in a porous journal bearing is simplified to a channel with uniform flow, which is partially filled with a porous medium. The simplified problem has zero fluid velocity q_x at the channel boundaries, moreover is the bottom boundary assumed sufficiently far away (i.e. outside of the figure) such that it has no influence on the fluid flow at the interface.

The original approach from Morgan et al. [106] for porous journal bearings ignored any slip effect at the porous-fluid interface. This means that they assumed the standard boundary conditions which also are applied for the normal Reynolds equation (the reader is referred to appendix A for a detailed description on the application of these boundary conditions). Figure 2.6a shows the corresponding fluid velocity field. It is obvious that an error is made when applying this approach.

To correct for this several adjustments can be found in literature. The most simple solution is to demand that the fluid velocity has to be continuous. An application of this condition for porous journal bearings is found in literature [103] and it is illustrated in figure 2.6b. It is pointed out by Meurisse et al. [103] that for practical applications this boundary condition has little to no influence on the fluid film pressure in comparison with the simplified approach from figure 2.6a.

Still, a discontinuity is observed in the fluid flow derivative in figure 2.6b. Corrections for this were proposed by Beavers et al. [9] and Brinkman [20], respectively shown in figure 2.6c and figure 2.6d. It is clear that the latter produces the most satisfying result, but also will increase the problem complexity. The correction shown in figure 2.6c is applied more easily and it can be argued that – at least for the fluid film flow – also produces the correct behavior. In fact it has been shown that when the so-called slip-coefficient α of the BEAVERS/JOSEPH model is chosen such that $\tilde{\alpha} = \sqrt{\tilde{\eta}/\eta}$ (where $\tilde{\eta} = \eta/\bar{f}$ is a parameter in the BRINKMAN model), both models produce the same velocity profile in the fluid film [116]. This also introduces the problem of parameter determination. Since these boundary conditions were only studied for the current problem formulation and not for the application in this thesis, finding the right parameter $\tilde{\alpha}$ or $\tilde{\eta}$ proves to be difficult and elaborate. Moreover, it has been pointed out in literature that even these more complex models have little influence on the essential behavior of porous journal bearings [103]. Therefore, the use of these porous-fluid interface models will be omitted in this work and the assumptions made in figure 2.6a are adopted.

Chapter 3

Dynamic systems containing revolute joints with clearance

In this chapter the kinematics and dynamics of revolute joints with clearance in the context of multibody systems are introduced. The representation of all necessary degrees of freedom in a plain journal bearing with misalignment in the proper reference frame is covered and the equation of motion for a misaligned plain journal bearing in the context of a rotor-bearing system is derived [157].

The rotor-bearing system containing revolute joints (i.e. plain journal bearings) that is under consideration in this work is a nonlinear dynamical system. In brief, the steady-state solutions and bifurcation points occurring in the considered examples as well as a short overview of the stability criteria for dynamical systems are given in this chapter [87, 142, 152]. These will be used later on to analyze the dynamical behavior of rotor-bearing systems containing plain journal bearings.

3.1   Multibody systems

In terms of a multibody system the plain journal bearing can be seen as a revolute joint with clearance, where the fluid film and asperity contacts generate forces and moments even though the nominal contours of the journal and bearing are not in contact. A conventional revolute joint only has one degree of freedom (DOF). The introduction of radial clearance leads to additional DOF: a total of three DOF for aligned journal bearings and five DOF for misaligned journal bearings when motion in axial direction is not considered. The kinematics of a plain journal

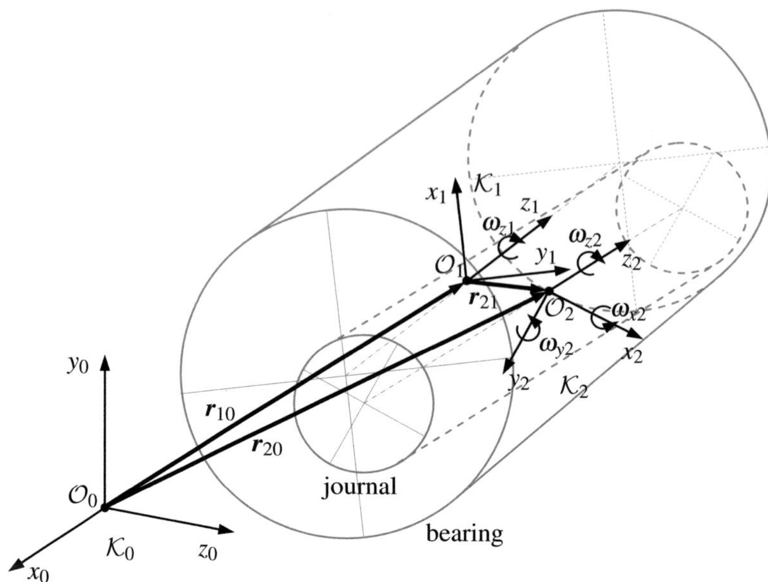

Figure 3.1: Orientation of the journal ($\mathcal{K}_2$) and the bearing ($\mathcal{K}_1$) of a plain journal bearing with respect to the reference frame $\mathcal{K}_0$.

bearing will be addressed by describing the motion of a misaligned journal with respect to the bearing's reference frame. These are used to properly include the motion according to the Reynolds equation. Following, the topic of dynamics is addressed by deriving the equations of motion for a misaligned plain journal bearing in which the kinematic relationships are used.

3.1.1 Kinematics

The relative (angular) displacements as well as the relative (angular) velocities of the journal and the bearing are needed to calculate forces and moments acting inside a plain journal bearing. For an aligned plain journal bearing where the bearing is fixed to the global reference frame and the journal's axis is parallel to the bearing's axis this is trivial. When the bearing's reference frame possesses a different motion with respect to the global reference frame (ground) and the orientation of the journal differs from that of the bearing extra efforts are needed to obtain the appropriate journal's motion. Figure 3.1 shows an arbitrary orientation of a misaligned plain journal bearing in a global reference frame $\mathcal{K}_0$. The bearing's reference frame is indicated with $\mathcal{K}_1$ and the journal's reference frame with $\mathcal{K}_2$. The representation of the motion of an arbitrary reference frame $\mathcal{K}_i$ relative to another reference frame (e.g. a global reference frame) $\mathcal{K}_j$ is a superposition of the translational motion of the origin $\mathcal{O}_i$ with respect to $\mathcal{O}_j$ and the rotational motion of $\mathcal{K}_i$ with respect to the reference frame $\mathcal{K}_j$. Such a translational motion can be deduced from the vector ${}^k\boldsymbol{r}_{ij}$ between $\mathcal{O}_i$ and $\mathcal{O}_j$, where the subscripts indicate both origin points and the preceding superscript k indicates the reference frame in which the quantity is expressed. Examples of such vectors are found in figure 3.1. This alone suffices to describe the motion in aligned plain journal bearings. Additionally, for misaligned plain journal bearings the rotational motion needs to be taken into account. The relative rotational motion between $\mathcal{K}_i$ and $\mathcal{K}_j$ can for example be described with the following three consecutive rotations:

$$\boldsymbol{T}_1(\phi_x) = \begin{bmatrix} 1 & 0 & 0 \\ 0 & \cos\phi_x & -\sin\phi_x \\ 0 & \sin\phi_x & \cos\phi_x \end{bmatrix} \quad \text{for} \quad \mathcal{K}_i \to \mathcal{K}_i', \tag{3.1}$$

$$\boldsymbol{T}_2(\phi_y) = \begin{bmatrix} \cos\phi_y & 0 & \sin\phi_y \\ 0 & 1 & 0 \\ -\sin\phi_y & 0 & \cos\phi_y \end{bmatrix} \quad \text{for} \quad \mathcal{K}_i' \to \mathcal{K}_i'' \tag{3.2}$$

and

$$\boldsymbol{T}_3(\phi_z) = \begin{bmatrix} \cos\phi_z & -\sin\phi_z & 0 \\ \sin\phi_z & \cos\phi_z & 0 \\ 0 & 0 & 1 \end{bmatrix} \quad \text{for} \quad \mathcal{K}_i'' \to \mathcal{K}_j, \tag{3.3}$$

which depend on the rotation parameters ϕ_x, ϕ_y and ϕ_z (these rotations always are about the axis in current coordinate system, e.g. the rotation ϕ_y is about the y-axis belonging to $\mathcal{K}_i'$). Depending on the simulation environment the order of the rotations and the choice of rotation axes may differ. Using these rotation matrices a vector can be represented by coordinates in different reference frames, e.g. ${}^j\boldsymbol{r} = \boldsymbol{T}_3\boldsymbol{T}_2\boldsymbol{T}_1{}^i\boldsymbol{r}$. Referring to figure 3.1 the rotations ϕ_x and ϕ_y correspond to the misalignment or tilting of the journal relative to the bearing and ϕ_z is the axial rotation of the journal. Under normal operating conditions for plain journal bearings, the former two rotations are orders of magnitude smaller than the latter one. To be able to describe the film thickness function and its time derivative for a misaligned plain journal bearing the (angular) displacements (e_x, e_y, ϕ_x and ϕ_y) and their respective time derivatives ($\dot{e}_x$, $\dot{e}_x$, $\dot{\phi}_x$ and $\dot{\phi}_y$) of the journal ($\mathcal{K}_2$) relative to the bearing's reference frame ($\mathcal{K}_1$) are needed. In addition, the angular velocity about the journal's axis Ω_2 and the bearing's axis Ω_1 are needed, which can be readily deduced by observing the angular velocities of the reference frames belonging to the journal ($\mathcal{K}_1$) and the bearing ($\mathcal{K}_2$). The relative displacements and velocities are described by ${}^1\boldsymbol{r}_{21}$ and ${}^1\dot{\boldsymbol{r}}_{21}$. Given the orientations of $\mathcal{K}_1$ and $\mathcal{K}_2$, the rotations

ϕ_x and ϕ_y are determined by the consecutive rotations from equation (3.1) and (3.2). The time derivatives of the angular coordinates of the journal in the bearing's reference frame $\mathcal{K}_1$ are given by the following kinematic differential equation [76]

$$
\begin{bmatrix} \dot{\phi}_x \\ \dot{\phi}_y \\ \dot{\phi}_z \end{bmatrix} = \frac{1}{\cos\phi_y} \begin{bmatrix} \cos\phi_z & -\sin\phi_z & 0 \\ \cos\phi_y\sin\phi_z & \cos\phi_y\cos\phi_z & 0 \\ -\sin\phi_y\cos\phi_z & \sin\phi_y\sin\phi_z & \cos\phi_y \end{bmatrix} \begin{bmatrix} \omega_{x2} \\ \omega_{y2} \\ \omega_{z2} \end{bmatrix},
\tag{3.4}
$$

where $\omega_{x2,y2,z2}$ are the coordinates of the angular velocity of the journal in (or relative to) the reference frame $\mathcal{K}_1$. It should be noted that here a singularity occurs when $\phi_y = \pi/2$. Fortunately, due to the fact that the clearance in journal bearings is very small compared to the journal radius $\phi_y \ll 1$ holds and therefore singularities will not occur. This also will allow for several simplification as will be shown in the next section. Here, equation (3.4) will be used to derive the equation of motion. This equation is also used to obtain the correct values of $\dot{\phi}_x$ and $\dot{\phi}_y$ in the context of a multibody software environment.

Reynolds equation

To properly include the translational and rotational velocities of the journal and the bearing in the Reynolds equation they need to be represented in the same reference frame. A close-up of the geometries of the journal and bearing surface for a misaligned configuration is shown in figure 3.2. The axis denotation and its relation with the velocities U, V and W in this figure is analogue to that of figure 2.1.

Consider two corresponding points; one on the journal's and one the bearing's surface that are separated by a film height H. These points possess a velocity described with the vector $\mathbf{v}_2$ for the journal and $\mathbf{v}_1$ for the bearing. The velocities include the translational motion as well as the angular velocity which results from the rotation about the bearing's and journal's axis (Ω_1 and Ω_2). Analogue to figure 3.1 a coordinate system $\mathcal{K}_1$ for the bearing is defined. This will be the reference frame in which the velocities appearing in the Reynolds equation for these two points are expressed. The journal's reference frame is indicated with $\mathcal{K}_2$. For a misaligned journal bearing the reference frame $\mathcal{K}_2$ is rotated with respect to $\mathcal{K}_1$ by the angles φ_x and φ_y due to misalignment and by φ_z due to a possible change in the local curvature of the surfaces at the corresponding points. It is assumed that the velocity components of the translational velocities are assumed small in comparison to the velocity resulting from the rotational motion. In addition, it is assumed that the angles which define the difference in curvature and rate of misalignment are small, i.e. $\varphi_x \ll 1$, $\varphi_y \ll 1$ and $\varphi_z \ll 1$. These are valid assumptions for the

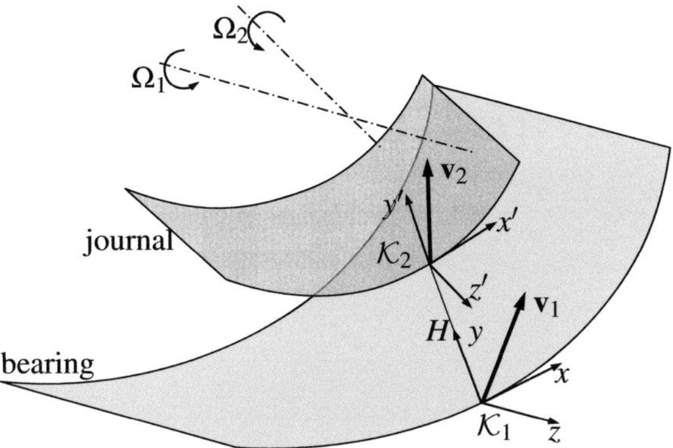

Figure 3.2: A close-up of the geometry of the journal's and bearing's surface for a misaligned configuration.

plain journal bearing, since the radial clearance is orders of magnitude smaller than the bearing radius and length. The point at the bearing's surface possesses the velocity vector

$$^1\mathbf{v}_1 = (v_{x1} + v_{R1})\mathbf{i}_x + v_{y1}\mathbf{i}_y + v_{z1}\mathbf{i}_z \tag{3.5}$$

defined in the reference frame $\mathcal{K}_1$ and the point at the journal's surface

$$^2\mathbf{v}_2 = (v_{x2} + v_{R2})\mathbf{i}_{x'} + v_{y1}\mathbf{i}_{y'} + v_{z1}\mathbf{i}_{z'}, \tag{3.6}$$

which is defined in the reference frame $\mathcal{K}_2$. Using the transformation matrices in equation (3.1)-(3.3) and assuming small angles, the projection of all velocity components in the reference frame $\mathcal{K}_1$ results in [38]

$$U_1 \approx v_{R1}, \quad V_1 = v_{y1} - V_\mathrm{p} \quad \text{and} \quad W_1 = v_{z1} \tag{3.7}$$

for the bearing and

$$U_2 \approx v_{x2} + v_{R2} - v_{y2}\varphi_z + \varphi_y v_{z2} \approx v_{R2}, \tag{3.8}$$

$$V_2 \approx (v_{x2} + v_{R2})\varphi_z + v_{y2} - \varphi_x v_{z2} \approx U_2\frac{\partial H}{\partial x} + v_{y2}, \tag{3.9}$$

$$W_2 \approx -(v_{x2} + v_{R2})\varphi_y + v_{y2}\varphi_x + v_{z2}, \tag{3.10}$$

for the journal. V_p represents the (optional) velocity corresponding to the fluid exchange between the fluid film and the porous medium, which will be defined later on in chapter 4. In this work motion in the axial direction of the plain journal bearing is not taken into account and terms containing the velocities $W_{1,2}$ are therefore ignored. Using these assumptions the right hand side of the Reynolds equation (2.1) can now be rewritten to [157]

$$H\frac{\partial}{\partial x}\frac{U_1+U_2}{2}+\frac{U_1-U_2}{2}\frac{\partial H}{\partial x}+H\frac{\partial}{\partial z}\frac{W_1+W_2}{2}+\frac{W_1-W_2}{2}\frac{\partial H}{\partial z}+(V_2-V_1)\approx\ldots$$
$$\ldots\frac{\partial}{\partial x}\left(H\frac{U_1+U_2}{2}\right)+\frac{\partial H}{\partial t}+V_{\mathrm{p}}.\tag{3.11}$$

3.1.2 Dynamics

Having obtained all model variables the equation of motion for a plain journal bearing can be constructed. These equations can be deduced from the Newton-Euler equations[1];

$$m\frac{_0\mathrm{d}^2}{\mathrm{d}t^2}\mathbf{r}_{20}=m\,_0\ddot{\mathbf{r}}_{20}=\sum_i\mathbf{F}_i\quad\text{and}\quad\frac{_0\mathrm{d}}{\mathrm{d}t}(\mathscr{I}\,\boldsymbol{\omega}_{20})=\sum_i\mathbf{M}_i,\tag{3.12}$$

where m indicates the journal's or rotor's mass, $\mathbf{F}_i$ indicate the forces and $\mathbf{M}_i$ the moments (or torques) acting on the journal and $\mathscr{I}$ refers to the moment of inertia tensor of the journal or rotor. It is noted that the center of mass of the rotor is here fixed at $\mathcal{O}_2$. To simplify matters the angular momentum is evaluated in the journal's reference frame $\mathcal{K}_2$. This is advantageous, because the journal's moment of inertia in this reference frame does not depend on time and is here taken as

$$^2\mathscr{I}=\begin{bmatrix}I_x&0&0\\0&I_y&0\\0&0&I_z\end{bmatrix},\tag{3.13}$$

where the values on the diagonal depend on the journal's geometrical shape. Using Euler's law the time derivative of the angular momentum becomes

$$\frac{_0\mathrm{d}}{\mathrm{d}t}(\mathscr{I}\,\boldsymbol{\omega}_{20})=\frac{_2\mathrm{d}}{\mathrm{d}t}(\mathscr{I}\,\boldsymbol{\omega}_{20})+\boldsymbol{\Lambda}_{0\to2}\times(\mathscr{I}\,\boldsymbol{\omega}_{20}),\tag{3.14}$$

where $\boldsymbol{\Lambda}_{0\to2}$ represents the angular velocity corresponding to the journal's orientation with respect to the global reference frame. The acceleration and angular velocity appearing in equation (3.12) can be deduced from the motion of $\mathcal{K}_1$ relative to $\mathcal{K}_0$ and the motion of $\mathcal{K}_2$ relative to $\mathcal{K}_1$:

$$_0\ddot{\mathbf{r}}_{20}=\,_0\ddot{\mathbf{r}}_{10}+\,_0\ddot{\mathbf{r}}_{21}\quad\text{and}\quad\boldsymbol{\omega}_{20}=\boldsymbol{\omega}_{10}+\boldsymbol{\omega}_{21}.\tag{3.15}$$

---

[1]A preceding subscript indicates the reference frame in which the time derivative is taken

Although in reality the bearings are often attached to the ground through some kind of supports, in this work the bearings are assumed to be fixed to the ground. With this simplification, the journal's acceleration simplifies to $_0\ddot{r}_{20} = _0\ddot{r}_{21}$ and its angular velocity to $\boldsymbol{\omega}_{20} = \boldsymbol{\omega}_{21}$.

The orientation of the journal with respect to the bearing can be realized by two consecutive rotations, defined here as ϕ_x followed by ϕ_y. To this end the rotation matrices from equation (3.1) and equation (3.2) are used. Using these transformations or the inverse of equation (3.4)

$$^{2}(\mathscr{I}\,\boldsymbol{\omega}_{20}) = I_x \cos\phi_y \dot{\phi}_x \mathbf{i}_x'' + I_y \dot{\phi}_y \mathbf{i}_x'' + I_z(\Omega_2 + \sin\phi_y \dot{\phi}_x)\mathbf{i}_z'' \tag{3.16}$$

and

$$\boldsymbol{\Lambda}_{0\to2} = \cos\phi_y \dot{\phi}_x \mathbf{i}_x'' + \dot{\phi}_y \mathbf{i}_x'' + \sin\phi_y \dot{\phi}_x \mathbf{i}_z'', \tag{3.17}$$

where $\mathbf{i}_{x,y,z}''$ indicate the base vectors of $\mathcal{K}_2$. Since $\phi_{x,y} \ll 1$ the simplifications $\sin\phi_{x,y} \approx \phi_{x,y}$ and $\cos\phi_{x,y} \approx 1$ are in order. With this equation (3.14) can be simplified to

$$\frac{_0\mathrm{d}}{\mathrm{d}t}{}^{2}(\mathscr{I}\,\boldsymbol{\omega}_{20}) = \begin{bmatrix} \cos\phi_y I_x \ddot{\phi}_x + \Omega_2 I_z \dot{\phi}_y + \sin\phi_y \dot{\phi}_x \dot{\phi}_y (I_z - I_y - I_x) \\ I_y \ddot{\phi}_y - \Omega_2 I_z \dot{\phi}_x + \sin\phi_x \cos\phi_x \dot{\phi}_x^2 (I_x - I_z) \\ \sin\phi_y I_z \ddot{\phi}_x + \cos\phi_y \dot{\phi}_x \dot{\phi}_y (I_z + I_y - I_x) \end{bmatrix} \approx \begin{bmatrix} I_x \ddot{\phi}_x + \Omega_2 I_z \dot{\phi}_y \\ I_y \ddot{\phi}_y - \Omega_2 I_z \dot{\phi}_x \\ \phi_y I_z \ddot{\phi}_x + \dot{\phi}_x \dot{\phi}_y (I_z + I_y - I_x) \end{bmatrix} \tag{3.18}$$

where the third term of the x- and y-component can be ignored because $\phi_{x,y} \ll 1$. Another implication of this is that the order of the rotations ϕ_x and ϕ_y is not important and will not affect the result. This work is not concerned with the axial moments and forces and therefore the conservation of angular momentum in the z-direction is not considered beyond this point. This leads to the following equation of motion

$$m\ddot{e}_x = F_x + F_{\mathrm{a},x} \tag{3.19}$$

$$m\ddot{e}_y = F_y + F_{\mathrm{a},y} \tag{3.20}$$

$$I_x \ddot{\phi}_x + I_z \dot{\phi}_y \Omega_2 = M_x + M_{\mathrm{a},x} \tag{3.21}$$

$$I_y \ddot{\phi}_y - I_z \dot{\phi}_x \Omega_2 = M_y + M_{\mathrm{a},y}, \tag{3.22}$$

where $F_{x,y}$ are the bearing forces and $M_{x,y}$ the bearing moments. An additional subscript a indicates an externally applied force or moment.

3.2 Nonlinear dynamics

The equation of motion which was derived in the previous section is a nonlinear dynamic system, since the bearing forces and moments are nonlinear. By formulating it with the following set of equations

$$
\begin{cases}
\dot{u}_1 = \dot{e}_x \\[4pt]
\dot{u}_2 = (F_x + F_{a,x})/m \\[4pt]
\dot{u}_3 = \dot{e}_y \\[4pt]
\dot{u}_4 = (F_y + F_{a,y})/m \\[4pt]
\dot{u}_5 = \dot{\phi}_x \\[4pt]
\dot{u}_6 = (M_x + M_{a,x} - I_z \dot{\phi}_y \Omega_2)/I_x \\[4pt]
\dot{u}_7 = \dot{\phi}_y \\[4pt]
\dot{u}_8 = (M_y + M_{a,y} + I_z \dot{\phi}_x \Omega_2)/I_y
\end{cases}
\tag{3.23}
$$

the dynamic system is described with a first order differential equation in the following state space form

$$
\dot{u} = f(u, \lambda),
\tag{3.24}
$$

which is an autonomous system since the vector field $f \in \mathbb{R}^n$ does not explicitly depend on time (t). In this equation $u(t) \in \mathbb{R}^n$ is the state vector and $\lambda \in \mathbb{R}$ is a bifurcation parameter. In this thesis this mainly will be the sum angular frequency of the journal and the bearing in a plain journal bearing (other parameters are also varied, but are not taken as bifurcation parameters). The state vector will contain – among other quantities – the (angular) displacements and (angular) velocities of the journal's center at the mid-plane with respect to the bearing's center at the mid-plane.

For a given starting point $u(t = 0) = u_0$ the dynamical system may follow different trajectories depending on the initial conditions and the system parameters. The main concern is how these solutions will look like in the long run. Will it converge to an equilibrium point or limit cycle? A fixed-point solution is a steady-state solution corresponding to an equilibrium position of the dynamical system for which the time derivative of the state vector is defined to be zero. When this is the case fixed-points u^0 abide $f(u^0, \lambda) = 0$. An example for a fixed-point solution for a rotor-bearing system is shown in figure 3.3a.

Periodic solutions return to the same state vector after a certain minimum period $T > 0$, i.e. $u^P(t + T, u_0) = u^P(t, u_0) \ \forall \ T \in \mathbb{R}$. An example of a periodic solution is shown in figure 3.3b. There are two more kinds of solutions that can occur for the studied dynamical systems. These

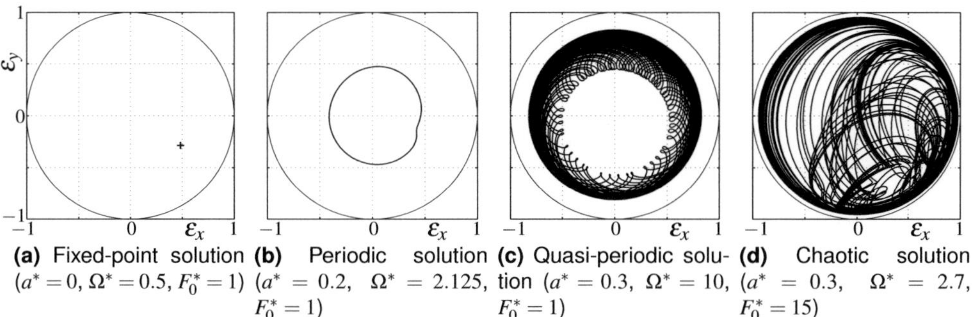

(a) Fixed-point solution $(a^* = 0, \Omega^* = 0.5, F_0^* = 1)$ **(b)** Periodic solution $(a^* = 0.2, \Omega^* = 2.125, F_0^* = 1)$ **(c)** Quasi-periodic solution $(a^* = 0.3, \Omega^* = 10, F_0^* = 1)$ **(d)** Chaotic solution $(a^* = 0.3, \Omega^* = 2.7, F_0^* = 15)$

Figure 3.3: Examples of possible orbits for an elementary rotor-bearing system without bearing misalignment. The dynamical system and its corresponding parameters are introduced in chapter 6.

are quasi-periodic and chaotic solutions. The detection of chaotic and quasi-periodic solutions is still an ongoing field of research and therefore most path-following algorithms are not able to identify them. Because of this, the consideration of quasi-periodic and chaotic solutions is beyond the scope of this work. An example for a quasi-periodic and chaotic solution are given in figures 3.3c and 3.3d. The interested reader is referred to e.g. [1, 122] for other examples of such solutions.

Using a path-following algorithm only fixed-point and periodic solutions of systems of equations can be followed when certain system parameters are varied. This method is often also referred to as numerical continuation.

3.2.1 Stability

The following lists criteria to determine the stability of fixed-point and periodic solutions. When considering a fixed-point solution of the dynamical system in equation (3.24), it is called stable when all nearby trajectories remain close to it for all time. In mathematical terms a solution $u^0(t)$ is considered stable in the sense of LYAPUNOV if for a perturbed solution $u(t)$ for any $\varepsilon > 0$ there exists a number $\delta = \delta(\varepsilon) > 0$ such that $\|u(t) - u^0(t)\| < \varepsilon$ for all $t \geq t_0$ if $\|u(t_0) - u^0(t_0)\| < \delta(\varepsilon)$.

Now assume that the dynamical system has a fixed-point solution u^0 which also is its initial condition: $u(0) = u^0$. After adding a small initial perturbation Δu to the solution $u(t) = u^0 + \Delta u$, substitution into equation (3.24) and a subsequent Taylor series expansion of f yields

$$\dot{u}^0 + \Delta \dot{u} = f(u^0 + \Delta u, \lambda) = f(u^0, \lambda) + \left.\frac{\partial f}{\partial u}\right|_{u=u^0} \Delta u + \mathcal{O}(|\Delta u|^2), \tag{3.25}$$

where it is assumed that f is sufficiently differentiable. When ruling out $\ddot{u}^0$ against $f(u^0, \lambda)$ and ignoring the higher order derivatives $\mathcal{O}(|\Delta u|^2)$ this results in the linearization of equation (3.24) [142]

$$\Delta \ddot{u} = J(u^0, \lambda)\Delta u, \tag{3.26}$$

where the Jacobian matrix is defined by $J = \frac{\partial f}{\partial u}\big|_{u=u^0}$. By observing its eigenvalues for a given fixed-point solution the stability of this solution can be determined. It is stable when all eigenvalues have negative real parts and it is unstable when an eigenvalue exists with a real part greater than zero. A critical case exists when the largest eigenvalue has a real part which equals zero and then no statement can be given on the fixed-point solution's stability using the current linearization. If the linearized system in equation (3.26) is stable, then for a fixed point solution u^0 it can be proven that u^0 also is stable solution of the nonlinear system in equation (3.24) [67].

The stability of periodic solutions is determined using Floquet theory. Assume the dynamical system has a periodical solution u^P with period T. For a small initial perturbation $u(t) = u^P + \Delta u$ (analogue to the approach in equation (3.25)) linearization now yields

$$\Delta \dot{u}(t) = J(u^P, \lambda)\Delta u(t). \tag{3.27}$$

Although being time-variant this is a linear differential equation. Hence, solutions may be stated as

$$u(t) = \Phi(t)u(0) \tag{3.28}$$

using the so-called fundamental matrix $\Phi(t)$. Differentiation and substitution into equation (3.27) reveals

$$\dot{\Phi}(t) = J(u^P, \lambda)\Phi(t) \tag{3.29}$$

which has the initial condition $\Phi(0) = I$, where I is the identity matrix. From equation (3.28) it follows that

$$\Delta u(t + T) = \Phi(T)\Delta u(t). \tag{3.30}$$

Using the modal matrix R of right the eigenvectors of Φ, one can write $\Delta u = R\alpha$. With $L^\top R = I$ and $L^\top \Phi R = \Lambda = \mathrm{diag}(\lambda_i)$ being the diagonal matrix of eigenvalues of Φ and L the modal matrix of left eigenvectors, one obtains

$$\alpha_i(t + T) = \lambda_i(T)\alpha_i(t). \tag{3.31}$$

Obviously, the eigenvalues of $\boldsymbol{\Phi}(T)$ determine whether the components of $\Delta\boldsymbol{u}(0)$ along the eigenvectors will grow or decay over one period. Thus, to ascertain the stability of the periodic solution one observes the eigenvalues of the monodromy matrix $\boldsymbol{\Phi}(T)$, which are called Floquet multipliers. One Floquet multiplier always exists with magnitude one, the other multipliers can be used to determine the periodic solution's stability. The solution is stable when all of the remaining multipliers have a magnitude less than one. When a multiplier exists whose magnitude exceeds one the solution is unstable. A borderline case exists when two or more multipliers have the magnitude one and the remaining multipliers have a magnitude less than one: again, this case cannot be judged using the linearized equations.

3.2.2 Bifurcations

When a bifurcation parameter (here λ) crosses a critical value λ^* the nature of a steady-state solution can change significantly at bifurcation points. Knowing the location and the implications of bifurcation points is therefore useful when studying the dynamical behavior of rotor-bearing systems. The following treats fold and Hopf bifurcation which occur for fixed-point solutions and fold, torus and flip bifurcations which can occur for periodic solutions [152].

Fixed-point solutions

A fold (or saddle-node) bifurcation constitutes the transition from a coexisting stable and unstable fixed-point solution to a parameter range where no fixed-point solutions exist (at least locally). At the bifurcation point the fixed-point solutions coincide. Here, at least one of the eigenvalues will equal zero. Figure 3.4a illustrates a fold bifurcation.

A Hopf bifurcation occurs at the transition from a fixed-point solution to a periodic and coexisting fixed-point solution. At the transition the fixed-point and periodic solution coincide and the fixed-point has a purely imaginary pair of conjugated eigenvalues. Hopf bifurcations can appear in three different forms: super-critical, sub-critical and degenerate Hopf bifurcations. Examples of super- and sub-critical Hopf bifurcations are found in figures 3.4b and 3.4c. At a super-critical Hopf bifurcation a stable periodic solution will coincide with an unstable fixed-point solution, after which it continues as a stable fixed-point solution. Before the bifurcation point the periodic solution coexists with the fixed-point solution. The real parts of previously addressed fixed-point's eigenvalues now change sign from negative to positive. The opposite case is a sub-critical Hopf bifurcation where all solution branches have the opposite stability. Here the real parts of the pair of eigenvalues change sign from positive to negative. A degen-

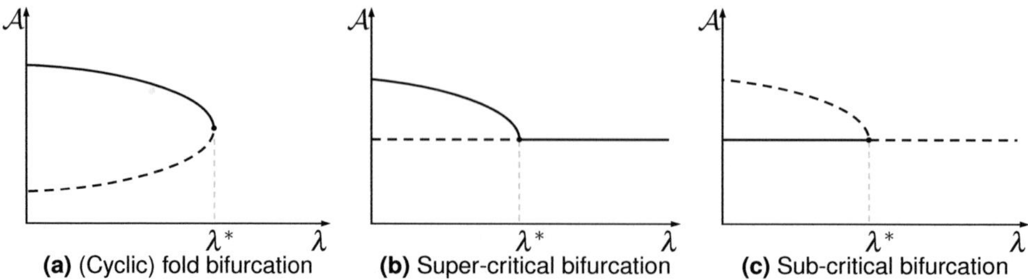

(a) (Cyclic) fold bifurcation   **(b)** Super-critical bifurcation   **(c)** Sub-critical bifurcation

Figure 3.4: Bifurcation diagrams with different kinds of bifurcations points. The scalar $\mathcal{A}$ is a general measure (chosen accordingly to the kind of solution), for plain journal bearings this can be the maximum eccentricity ratio for instance. Solid lines represent stable solutions and dashed lines represent unstable solutions.

erate Hopf bifurcation has no periodic solution at either side of the bifurcation and can be seen as borderline case.

Periodic solutions

The cyclic fold bifurcation for periodic solutions is analogue to the fold bifurcation for fixed-point solutions. Now, however, before the bifurcation point a stable and an unstable periodic solution coexist whereas after the transition no periodic solutions exist (at least locally). One (additional) Floquet multiplier with a magnitude greater than one appears when passing the bifurcation from the stable to the unstable solution (see figure 3.4a).

The torus, Neimark-Sacker or secondary Hopf bifurcation is in turn comparable to the normal Hopf bifurcation. This bifurcation constitutes the transition of a coexisting periodic and quasi-periodic solution to a periodic solution. At the bifurcation point the quasi-periodic and periodic solution coincide and the periodic solution has a complex conjugated pair of Floquet multipliers with magnitude one. The multipliers move into the unit circle of the complex plane for a super-critical torus bifurcation whereas for a sub-critical bifurcation the opposite happens. When proper scalar measures are chosen to represent quasi-periodic and periodic solutions figures 3.4c and 3.4b can also illustrate these bifurcations, although it should be noted that it might not be possible to find a proper scalar measure for qausi-periodic solutions.

Finally, the flip bifurcation or period doubling bifurcation constitutes the transition of the coexistence of a period-n and period-$2n$ solution to a period-n solution, where $n \in \mathbb{N}$. At the flip bifurcation point the period-n and period-$2n$ solution coincide, where the n-period solution has a multiplier which equals -1. The flip bifurcation is super-critical when a Floquet multiplier moves into the unit circle of the complex plane, whereas it is sub-critical when a multiplier moves outside of the unit circle. Again by choosing proper scalar measures figures 3.4c and 3.4b can also illustrate these bifurcations.

Chapter 4

Models for plain journal bearings

This chapter treats several models to calculate the forces and moments generated in plain journal bearings. First, the governing equations and their accompanying boundary conditions as well as definitions for the bearing forces and moments are given. After that analytical, semi-analytical and numerical models for solid and porous journal bearings are introduced.

4.1 The plain journal bearing

The top-left panel in figure 4.1 shows the schematic cross-section at the mid-plane ($z = 0$) of a plain journal bearing with an arbitrary eccentricity. The origin of the x, y-coordinate system is positioned in the center of the bearing. The journal rotates with angular frequency Ω_2 around its own axis and its center is displaced relative to the center of the bearing with an eccentricity vector e at the mid-plane. The bearing has an inner radius R_i and an outer radius R_o and rotates with an angular frequency Ω_1 around its axis. The attitude angle γ is defined as the angle between the eccentricity vector and the y-axis. The remaining panels in figure 4.1 show the geometry in a plain journal bearing when the journal axis is misaligned with respect to the bearing axis. Next to an eccentric displacement at the mid-plane the journal axis is tilted by ϕ_x and ϕ_y around the x- and y-axis respectively. The journal has a radius R and its difference compared to the bearing inner radius is denoted with the radial clearance c.

The film thickness is expressed as a function of the axial coordinate z and the circumferential coordinate θ and will be derived in the following. Observe the triangle ABC in the bottom-right panel in figure 4.1. Here, C is an arbitrary point along the journal axis, which depends on the

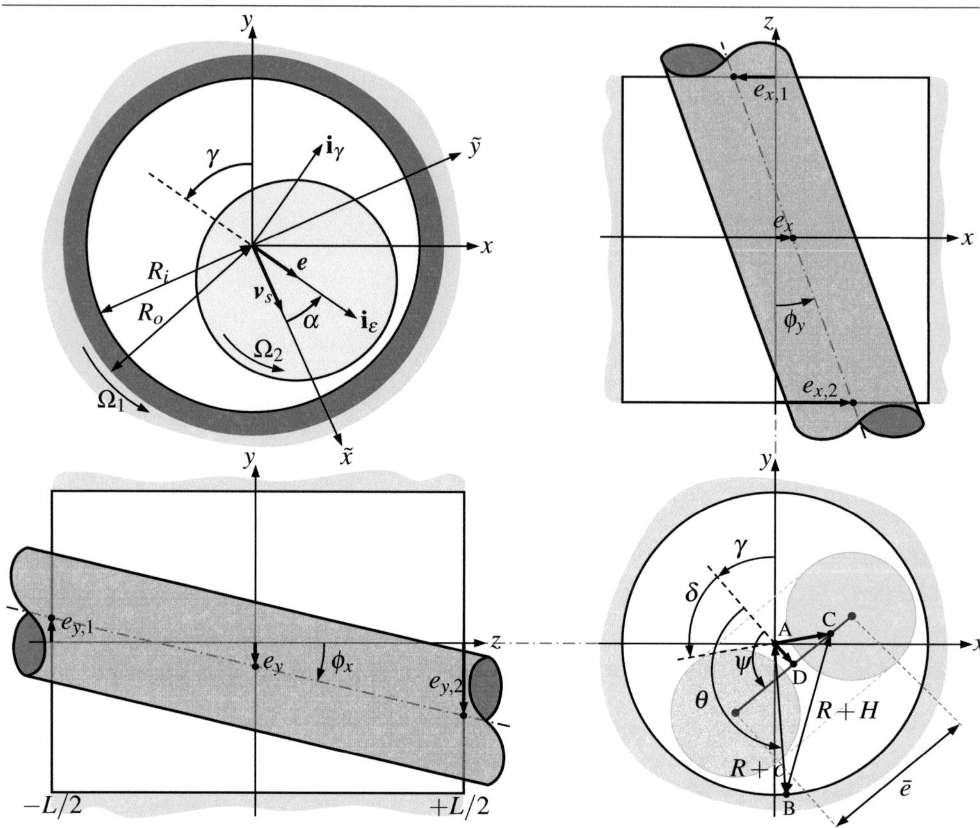

Figure 4.1: Geometry of a plain journal bearing. The top-left panel shows a cross-section with geometrical parameters including the pure squeeze velocity vector. The remaining panels describe misalignment in a plain journal bearing.

axial coordinate. By applying the law of cosines and ignoring the second order terms in H/R, c/R and $\overline{AC}/R$ it can be seen that [21]

$$(R+H)^2 = (R+c)^2 + \overline{AC}^2 + 2(R+c)\overline{AC}\cos(\theta-\delta) \quad \Leftrightarrow \quad H \approx c + \overline{AC}\cos(\theta-\delta). \quad (4.1)$$

Now observe the triangle ADC, where D located on the journal axis at the mid-plane. Here, $\overline{AD} = e \equiv \|\boldsymbol{e}\|$ and $\overline{CD} = \bar{e}z/L$, where $\bar{e}$ is defined as the length of the journal axis projected on the mid-plane. For this triangle the law of cosines and the law of sines respectively result in

$$\overline{AC}\cos\delta = e + \frac{z}{L}\bar{e}\cos\psi \quad \text{and} \quad \overline{AC}\sin\delta = \frac{z}{L}\bar{e}\sin\psi,$$

where ψ is defined as the angle between the eccentricity vector and the projection of the journal axis on the mid-plane. When combined with equation (4.1) this results in the film thickness function for a misaligned plain journal bearing

$$H(\theta,z) = c + e\cos\theta + \frac{z}{L}\bar{e}\cos(\theta - \psi) \quad \text{or} \quad h(\theta,z^*) = 1 + \varepsilon\cos\theta + z^*\bar{\varepsilon}\cos(\theta - \psi), \quad (4.2)$$

where $h = H/c$ is the dimensionless film thickness function, $\varepsilon = e/c$ is the dimensionless eccentricity ratio (at the mid-plane), $z^* = z/L$ is the dimensionless axial coordinate and $\bar{\varepsilon} = \bar{e}/c$ is the dimensionless projection of the journal axis on the mid-plane. When no misalignment occurs, i.e. $\bar{\varepsilon} = 0$, this function reduces to classical film thickness function for an aligned plain journal bearing; $h(\theta) = 1 + \varepsilon\cos\theta$. The reader is referred to appendix B for a list of the derived parameters ε, $\bar{\varepsilon}$, γ and ψ and their dependency on the input parameters ε_x, ε_y, ϕ_x and ϕ_y (see figure 4.1).

The Reynolds equation introduced in equation (2.1) is used to describe the pressure in the fluid film of the plain journal bearing. Next to the assumptions made in chapter 2, it is assumed that the journal's motion with respect to the bearing in the axial direction can be ignored.

A porous journal bearing distinguishes itself from a solid journal bearing due to its bush which possesses a certain porosity. This porous bush can contain lubricant and therefore also sustain a pressure field, which gives rise to a rate of flow per unit area between the film and the bush. This flow is introduced through the third term $(V_2 - V_1)$ on the right hand side of the Reynolds equation (2.1). This term represents the change in film thickness over time but also the velocity of the fluid particles entering or leaving the fluid film from or to the porous bush as was explained in section 3.1.1. The latter velocity can directly be determined with Darcy's law:

$$V_{\mathrm{p}} = \frac{K}{\eta}\left[\frac{\partial\widehat{p}_{\mathrm{h}}}{\partial r}\right]_{r=R_i}. \quad (4.3)$$

More details on the velocity boundary conditions for the Reynolds equation are found in appendix A. Velocity conditions in axial and circumferential direction have been omitted since they have no significant influence for porous journal bearings as was explained in section 2.2.2. The dimensionless formulation from [152] is adopted for the pressure and time respectively:

$$p_{\mathrm{h}}^* = (c/R_i)^2 p_{\mathrm{h}}/6\eta\Omega_0 \quad \text{and} \quad \tau = \Omega_0 t, \quad (4.4)$$

where Ω_0 is a parameter to scale the angular frequency. This parameter will be defined later on in chapter 6. With this the Reynolds equation can be written as

$$\frac{\partial}{\partial\theta}\left(h^3\frac{\partial p_{\mathrm{h}}^*}{\partial\theta}\right)+\kappa^2\frac{\partial}{\partial z^*}\left(h^3\frac{\partial p_{\mathrm{h}}^*}{\partial z^*}\right)+12\Psi\left[\frac{\partial\widehat{p}_{\mathrm{h}}^*}{\partial r^*}\right]_{r^*=1}=\Omega^*\frac{\partial h}{\partial\theta}+2\frac{\partial h}{\partial\tau},\tag{4.5}$$

where $\kappa=R_i/L$, L is the bearing length and $\Omega^*=(\Omega_1+\Omega_2)/\Omega_0$. This equation is valid for porous and solid journal bearings. By choosing $\Psi=0$ the Reynolds equation returns to its original form which describes a solid bearing.

The fluid flow in the porous bush is governed by Darcy's law. This equation combined with the continuity equation was introduced in chapter 2 in equation (2.21). When assuming an isotropic, homogeneous permeability distribution, ignoring gravity effects and making the same assumptions as for equation (4.5) this equation now reads

$$\frac{\partial^2\widehat{p}_{\mathrm{h}}^*}{\partial r^{*2}}+\frac{1}{r^*}\frac{\partial\widehat{p}_{\mathrm{h}}^*}{\partial r^*}+\frac{1}{r^{*2}}\frac{\partial^2\widehat{p}_{\mathrm{h}}^*}{\partial\theta^2}+\kappa^2\frac{\partial^2\widehat{p}_{\mathrm{h}}^*}{\partial z^{*2}}=0,\tag{4.6}$$

when it is expressed in cylindrical coordinates. $\widehat{p}_{\mathrm{h}}^*$ represents the dimensionless pressure in the porous bush (using the same definition as for the dimensionless pressure p_{h}^* in the fluid film) and the dimensionless radial coordinate is defined as $r^*=r/R_i$. In using Darcy's law it is assumed that flow inside the porous medium is laminar and shearing and inertial effects can be ignored due to the fact that the seepage velocity as well as the corresponding velocity gradient are sufficiently small [79, 117].

Both equations are coupled through the third term on the left hand side of equation (4.5) and the continuity of pressure. The boundary conditions in axial and circumferential direction are

$$\widehat{p}_{\mathrm{h}}^*(\theta,z^*=-1/2,r^*)=\widehat{p}_{\mathrm{h}}^*(\theta,z^*=+1/2,r^*)=0,\quad\widehat{p}_{\mathrm{h}}^*(\theta=0,z^*,r^*)=\widehat{p}_{\mathrm{h}}^*(\theta=2\pi,z^*,r^*),\tag{4.7}$$

which also are valid for the Reynolds equation due to the first boundary condition in radial direction

$$p_{\mathrm{h}}^*(\theta,z^*)=\widehat{p}_{\mathrm{h}}^*(\theta,z^*,r^*=1),\tag{4.8}$$

which assures pressure continuity between the fluid film and the porous medium. At the outer bearing radius either

$$\left.\frac{\partial\widehat{p}_{\mathrm{h}}^*(\theta,z^*,r^*)}{\partial r^*}\right|_{r^*=R_o/R_i}=0\quad\text{or}\quad\widehat{p}_{\mathrm{h}}^*(\theta,z^*,r^*=R_o/R_i)=0\tag{4.9}$$

can be prescribed. These boundary conditions assume the porous bearing to be either embedded in a solid material or to be surrounded by ambient pressure, which are respectively a Neumann and a Dirichlet boundary condition. Unless stated otherwise, the Neumann boundary condition (the first boundary condition in equation (4.9)) will be used throughout this work. Please refer to figure 4.1 for the used parameters and variables.

4.1.1 Plain bearing forces and moments

Once the hydrodynamic pressure p_h^* (given by equation (4.5)) and the pressure arising from asperity contacts p_e^* (given by equation (2.13) or (2.14)) are known the force and moment components arising in the plain journal bearing can be evaluated by integration of the pressure function. Depending on the cavitation algorithm negative pressures may arise. These are excluded by integrating over the positive pressure range, because the cavitation pressure in the cavitation range (with negative pressure) is very small and can be assumed to be zero. The hydrodynamic dimensionless force components in the directions $\mathbf{i}_\varepsilon$ and $\mathbf{i}_\gamma$ (see the top left panel of figure 4.1) for the fluid film read

$$F_{h,\varepsilon}^* = 3 \int_{-1/2}^{+1/2} \int_{\theta_1}^{\theta_2} p_h^* \cos\theta \, d\theta \, dz^* \quad \text{and} \quad F_{h,\gamma}^* = 3 \int_{-1/2}^{+1/2} \int_{\theta_1}^{\theta_2} p_h^* \sin\theta \, d\theta \, dz^*, \tag{4.10}$$

where the dimensionless force is defined as $F_{(k,j)}^* = (c/R_i)^2 F_{(k,j)}/\eta\Omega_0 LD$ (where $k = \text{h}, \text{e}$), D denotes the inner bearing diameter and θ_1 and θ_2 respectively mark the beginning and end of the positive range of the pressure profile p_h^*. The magnitude of the hydrodynamical dimensionless load capacity is calculated with $F_h^* = \sqrt{F_{h,\varepsilon}^{*2} + F_{h,\gamma}^{*2}}$. For the asperity contacts the dimensionless force components read

$$F_{e,\varepsilon}^* = 3 \int_{-1/2}^{+1/2} \int_{0}^{2\pi} p_e^* \cos\theta \, d\theta \, dz^* \quad \text{and} \quad F_{e,\gamma}^* = 3 \int_{-1/2}^{+1/2} \int_{0}^{2\pi} p_e^* \sin\theta \, d\theta \, dz^*. \tag{4.11}$$

Using the definition of p_e from equation (2.13) or (2.14) the dimensionless contact pressure is written as $p_e^* = c_{GW}^{sr} F_{3/2}(h(\theta)c/\sigma_e)$ or $p_e^* = c_{GW}^{rr} F_{5/2}(h(\theta)c/\sigma_e)$ where

$$c_{GW}^{i} = \begin{cases} \dfrac{\sqrt{2}}{9}E\dfrac{\hat{\eta}\hat{\beta}\sigma_e}{\eta\Omega_0}\left(\dfrac{c}{R_i}\right)^2\sqrt{\dfrac{\sigma_e}{\hat{\beta}}} & \text{for} \quad i = sr \\[3mm] \dfrac{8\sqrt{2}}{45}\pi E\dfrac{(\hat{\eta}\hat{\beta}\sigma_e)^2}{\eta\Omega_0}\left(\dfrac{c}{R_i}\right)^2\sqrt{\dfrac{\sigma_e}{\hat{\beta}}} & \text{for} \quad i = rr \end{cases} \tag{4.12}$$

The parameter c_{GW}^{sr} is used for the smooth-rough contact and c_{GW}^{rr} is used for the rough-rough contact. These forces are added to the hydrodynamical force components to obtain the total force component; $F_{t,i}^* = F_{h,i}^* + F_{e,i}^*$ where $i = \varepsilon, \gamma$.

Analogously, the bearing moment components for the fluid film are calculated with

$$M_{h,\varepsilon}^* = -3\int_{-1/2}^{+1/2}\int_{\theta_1}^{\theta_2} p_h^* \sin\theta z^* d\theta dz^* \quad \text{and} \quad M_{h,\gamma}^* = +3\int_{-1/2}^{+1/2}\int_{\theta_1}^{\theta_2} p_h^* \cos\theta z^* d\theta dz^*, \tag{4.13}$$

where the dimensionless moment is defined as $M_{k,j}^* = (c/R_i)^2 M_{k,j}/\eta\Omega_0 L^2 D$ (where $k = h, e$). The total dimensionless hydrodynamic moment capacity is calculated with $M_h^* = \sqrt{M_{h,\varepsilon}^{*2} + M_{h,\gamma}^{*2}}$. For the asperity contacts the dimensionless moment components read

$$M_{e,\varepsilon}^* = -3\int_{-1/2}^{+1/2}\int_0^{2\pi} p_e^* \sin\theta z^* d\theta dz^* \quad \text{and} \quad M_{e,\gamma}^* = +3\int_{-1/2}^{+1/2}\int_0^{2\pi} p_e^* \cos\theta z^* d\theta dz^*. \tag{4.14}$$

These moments are added to the hydrodynamical moments to obtain the total moment component; $M_{t,i}^* = M_{h,i}^* + M_{e,i}^*$ where $i = \varepsilon, \gamma$. Finally, the force and moment vectors need to be transformed to the x,y-coordinate system using the attitude angle γ:

$$\begin{aligned} F_{t,x}^* &= F_{t,\varepsilon}^* \sin\gamma + F_{t,\gamma}^* \cos\gamma \\ F_{t,y}^* &= -F_{t,\varepsilon}^* \cos\gamma + F_{t,\gamma}^* \sin\gamma \end{aligned} \quad \text{and} \quad \begin{aligned} M_{t,x}^* &= M_{t,\varepsilon}^* \sin\gamma + M_{t,\gamma}^* \cos\gamma \\ M_{t,y}^* &= -M_{t,\varepsilon}^* \cos\gamma + M_{t,\gamma}^* \sin\gamma \end{aligned} \tag{4.15}$$

4.1.2 Friction force

The hydrodynamical friction force $F_{\mu,h}^*$ is found by integrating the viscous shear stress and the friction force $F_{\mu,e}^*$ due to asperity contacts by multiplying the normal contact force with the friction coefficient parameter μ_c[1]. When assuming a Newtonian fluid this eventually leads to [21]

$$F_{\mu,\mathrm{h}}^* = \frac{c}{2R_i} \int_{-1/2}^{+1/2} \int_0^{2\pi} \left(3h\frac{\partial p_\mathrm{h}^*}{\partial \theta} + \frac{\Omega_s^*}{h} \right) \mathrm{d}\theta \mathrm{d}z^* \quad \text{and} \quad F_{\mu,\mathrm{e}}^* = \mu_c \int_{-1/2}^{+1/2} \int_0^{2\pi} p_\mathrm{e}^* \cos\theta \mathrm{d}\theta \mathrm{d}z^*, \quad (4.16)$$

where $\Omega_s^* = (\Omega_2 - \Omega_1)/\Omega_0$. Assuming the film thickness function in equation (4.2) and accounting for the same cavitation range as for the force calculation the hydrodynamical friction force can be rewritten to

$$F_{\mu,\mathrm{h}}^* = \frac{c}{2R_i} \left(\varepsilon F_{\mathrm{h},\gamma}^* - \bar{\varepsilon} \left(\cos\psi M_{\mathrm{h},\varepsilon}^* + \sin\psi M_{\mathrm{h},\gamma}^* \right) + \Omega_s^* \int_{-1/2}^{+1/2} \left(\int_{\theta_1}^{\theta_2} \frac{\mathrm{d}\theta}{h} + \int_{\theta_2}^{\theta_1+2\pi} \frac{h(\theta_2)}{h^2} \mathrm{d}\theta \right) \mathrm{d}z^* \right).$$

$$(4.17)$$

For aligned journal bearings, the integrals on the right hand side can be solved analytically. These are so-called Sommerfeld integrals and the reader is referred to [21, 60] for a list of corresponding solutions. When considering misaligned bearings the integral in axial direction is approximated by a summation of terms with constant axial coordinate. This again allows for analytical integration in circumferential direction. The friction force for porous journal bearings does not include the viscous friction caused inside the cavitation area, which means that the integral from θ_2 to $\theta_1 + 2\pi$ in equation (4.17) is ignored. This assumption is justified by the observations made in [81, 105]. The inclusion of the friction forces in the bearing force and moment calculation has been omitted, since it has been observed that the friction force is insignificant compared to the bearing force components $F_{\mathrm{t},i}^*$. To compare the model with measurements the friction coefficient of the journal bearing is needed, which is defined as the ratio of the friction force and the total load capacity; $\mu = (F_{\mu,\mathrm{h}}^* + F_{\mu,\mathrm{e}}^*)/F_\mathrm{t}^*$.

4.1.3 Impedance method

Childs et al. [25] showed that the right hand side of the Reynolds equation can be rewritten for aligned cylindrical journal bearings by introducing a so-called *pure squeeze velocity vector*

$$v_s^* = \varepsilon' - (0,0,\tfrac{1}{2}\Omega^*) \times \varepsilon, \quad (4.18)$$

---

[1]The friction force in a journal bearing can be seen as a tangential force parallel to the journal's surface. In reality, the friction in a journal bearing is exercised as a moment about the journal's axis. Here, this moment would consist of the tangential friction force multiplied by the journal's radius.

where the $(\ )'$ indicates derivation with respect to τ, vectors in the x,y-coordinate system are indicated in bold and the dimensionless pure squeeze velocity vector is defined by $\boldsymbol{v}_s^* = \boldsymbol{v}_s/c\Omega_0$. When observing the motion of the journal's center in the rotating $\tilde{x},\tilde{y}$-coordinate system (see the top left panel in figure 4.1) it appears to be in a state of pure squeezing [25]. In adopting this formulation the right hand side of the Reynolds equation becomes

$$\Omega^*\frac{\partial h}{\partial \theta} + 2\frac{\partial h}{\partial \tau} = 2\left(\varepsilon'\cos\theta + \varepsilon(\gamma' - \tfrac{1}{2}\Omega^*)\sin\theta\right) = 2v_s^*\cos(\alpha+\theta), \tag{4.19}$$

where $v_s^* = \|\boldsymbol{v}_s^*\|$ and α indicates the angle between the pure squeeze velocity vector and the eccentricity vector. Due to the linear dependence of the Reynolds equation on the norm v_s^* of the pure squeeze velocity vector, it can be excluded from the Reynolds equation. In this case one solves for $\tilde{p}_{\mathrm{h}}^* = p_{\mathrm{h}}^*/v_s^*$. This does not influence the zero-crossings of the pressure function and also can be applied to porous bearings because Darcy's law also linearly depends on v_s^*. Instead of bearing forces one now calculates an impedance vector, defined by $\mathcal{W}_i^* = -F_{\mathrm{h},i}^*/v_s^*$. With this, look-up tables can be generated which are called impedance maps. When observing the impedance vector in the $\tilde{x},\tilde{y}$-coordinate system it depends only on two variables, namely ε and α for a given bearing geometry κ and permeability Ψ. In this work impedance maps are used to compare different plain journal bearing models. A description on how to calculate the modified pressure $\tilde{p}_{\mathrm{h}}^*$ with the proposed method will be introduced later on. Once $\mathcal{W}_\varepsilon^*$ and $\mathcal{W}_\gamma^*$ are obtained the impedance components in the $\tilde{x},\tilde{y}$-coordinate system for the impedance map are given by

$$\begin{aligned} \mathcal{W}_{\tilde{x}}^* &= \mathcal{W}_\varepsilon^*\sin\alpha + \mathcal{W}_\gamma^*\cos\alpha \\ \mathcal{W}_{\tilde{y}}^* &= \mathcal{W}_\varepsilon^*\cos\alpha - \mathcal{W}_\gamma^*\sin\alpha \end{aligned}. \tag{4.20}$$

When taking into account misalignment, the right hand side of the Reynolds equation is no longer linear in v_s^*. Because of this, it cannot be ruled out anymore by linear scaling of variables and the impedance method is therefore unusable. The right hand side is now written as

$$\Omega^*\frac{\partial h}{\partial \theta} + 2\frac{\partial h}{\partial \tau} = 2\left(\varepsilon'\cos\theta + \varepsilon(\gamma' - \tfrac{1}{2}\Omega^*)\sin\theta + z^*\left(\bar{\varepsilon}'\cos(\theta-\psi) + \bar{\varepsilon}(\gamma'+\psi' - \tfrac{1}{2}\Omega^*)\sin(\theta-\psi)\right)\right). \tag{4.21}$$

Apart from the eccentricity ratio ε and the attitude angle γ, parameters associated with the misalignment $\bar{\varepsilon}$ and ψ now also occur (see also equation (4.2)). This is the general case and will be used throughout the rest of this thesis. Expressions for the derived parameters depending on the input parameters ε_x, ε_y, ϕ_x and ϕ_y and their time derivatives ε_x', ε_y', ϕ_x' and ϕ_y' can be found in appendix B.

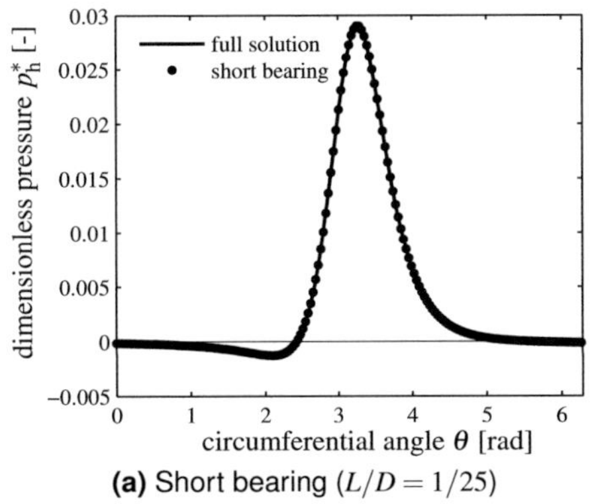

(a) Short bearing ($L/D = 1/25$)

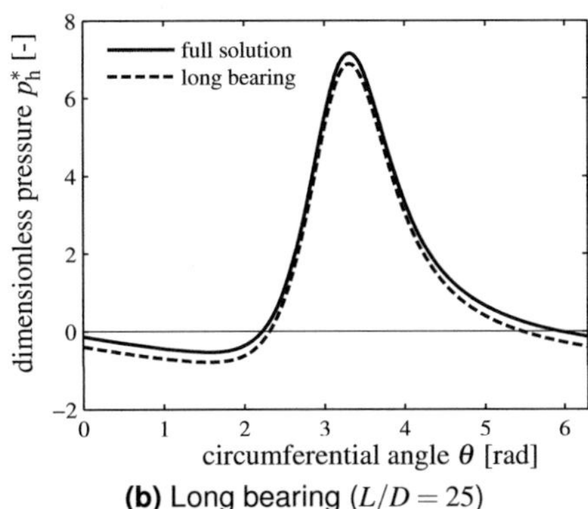

(b) Long bearing ($L/D = 25$)

Figure 4.2: Comparisons of the analytical short and long bearing pressure solutions with the pressure solution of the full Reynolds equation at $z^* = 0$. An arbitrary case with $\varepsilon_x = 0.5$, $\varepsilon'_x = 0.9$, $\varepsilon_y = -0.5$ and $\varepsilon'_y = 0.3$ is chosen here.

4.2 Analytical approximations

The journal bearing problem can be solved analytically for the long and short bearing simplifications as was shown e.g. in [25, 152]. For a short journal bearing ($L/D \lesssim 1/2$) the term $\partial p_h^*/\partial \theta$ can be ignored. Therefore, the first term on the left hand side of equation (4.5) vanishes and when taking the boundary conditions from equation (4.7) the pressure function becomes

$$p_h^*(\theta, z^*) = (z^{*2} - \tfrac{1}{4})(\varepsilon' \cos \theta + \varepsilon(\gamma' - \tfrac{1}{2}) \sin \theta)/\kappa^2 h^3, \tag{4.22}$$

which is positive between $\theta_1 = \tfrac{1}{2}\pi + \tan^{-1}\left(\varepsilon(\gamma' - \tfrac{1}{2}\Omega^*)/\varepsilon'\right)$ and $\theta_2 = \theta_1 + \pi$. This solution is also known as the OCVIRK solution [119].

For a long bearing ($L/D \gtrsim 2$), on the other hand, the term $\partial p_h^*/\partial z^*$ can be ignored and the second term on the left hand side of equation (4.5) vanishes. This solution also is known as the SOMMERFELD solution [144]. This combined with the boundary condition $p_h^*(\theta_1) = p(\theta_1 + \pi) = 0$ (as proposed by [60]) results in the pressure function

$$p_h^*(\theta) = \left(\varepsilon' \cos \theta + 2\varepsilon(\gamma' - \tfrac{1}{2}) \sin \theta/(2 + \varepsilon^2)\right)(2 + \varepsilon \cos \theta)/h^2, \tag{4.23}$$

which is positive between $\theta_1 = -\tan^{-1}\left(\varepsilon'(2 + \varepsilon^2)/2\varepsilon(\gamma' - \tfrac{1}{2}\Omega^*)\right)$ and $\theta_2 = \theta_1 + \pi$.

It should be noted that an error can be made when applying the long bearing solution. Whereas the short bearing solution converges to the solution of the full Reynolds equation when $L/D \to 0$

this is not the case for the long bearing solution when $L/D \to \infty$. First of all, an error is made at the axial boundaries where the long bearing solution cannot satisfy the prescribed boundary condition. For very long bearings, this error becomes insignificant. More worrying is the discrepancy observed when comparing the boundaries of the positive pressure range of both solutions. Although the shapes of the pressure profiles for a given axial coordinate seem identical a difference is observed. This can be explained by the condition $p_{\mathrm{h}}^*(\theta_1) = p_{\mathrm{h}}^*(\theta_1 + \pi) = 0$, which is needed to restrict the solution to a certain pressure height. Without this condition any valid solution with an added constant pressure would also be a valid solution. The condition itself is an assumption and apparently does not restrict the pressure function correctly. Figure 4.2 illustrates this.

To improve the quality of analytical solutions several corrections of the previously described analytical solutions or approximations of the full solution are found in literature. Both analytical functions can be combined to a finite bearing approximation [152]. Another approach is to use Galerkin's method to improve the quality of an analytical solution, where the analytical solution is used as ansatz function and the unaltered Reynolds equation as a residual function [137]. Other approximations can be found in literature, e.g. [4, 7]. However, all of these solutions remain approximations to the solution of the full Reynolds equation and extensions to these solutions to account for example for a porous bush, misalignment or other phenomena are problematic. To overcome these issues Galerkin's method is utilized in which the governing equation for the porous medium is used to construct an ansatz function for the pressure function.

4.3 An approximation using Galerkin's discretization method

In order to obtain a low-dimensional approximation Galerkin's method [136] with global ansatz functions is used to approximate the hydrodynamic pressure solution. To illustrate the origin of the used ansatz function the method is applied to the general case without misalignment ($h = 1 + \varepsilon \cos \theta$), but including the influence of a porous bush. Subsequently, the method will be extended to account for misalignment.

The governing equation for the fluid pressure inside the porous bush will be the starting point to obtain our ansatz function. Since for constant κ equation (4.6) is a Laplace equation, it can be solved by separation of variables [31]. By assuming that the solution can be written as a product of a circumferential, an axial and a radial solution,

$$\widehat{p}_{\mathrm{h}}^*(\theta, z^*, r^*) \cong O(\theta) Z(z^*) \rho(r^*), \tag{4.24}$$

the partial differential equation is separated into three ordinary differential equations,

$$\frac{\partial^2 O}{\partial \theta^2} + \lambda_1^2 O = 0, \tag{4.25}$$

$$\frac{\partial^2 Z}{\partial z^{*2}} + \lambda_2^2 / \kappa^2 Z = 0, \tag{4.26}$$

$$r^{*2} \frac{\partial^2 \rho}{\partial r^{*2}} + r^* \frac{\partial \rho}{\partial r^*} + (\lambda_1^2 - \lambda_2^2 r^{*2}) \rho = 0. \tag{4.27}$$

By applying the boundary conditions from equation (4.7) and (4.9) and the fact that the pressure profile must be symmetric in axial direction around $z^* = 0$ a solution for the pressure in the porous bush is obtained. When this solution is evaluated at the inner radius of the porous bush it can be used as an ansatz function to solve the Reynolds equation with Galerkin's method. With this the ansatz function for the fluid pressure reads

$$p_{\mathrm{h,a}}^*(\theta, z^*) = \widehat{p}_{\mathrm{h,a}}^*(\theta, z^*, r^* = 1) = \sum_{n,m} A_{nm}(r^* = 1)\big(a_{nm}\sin(n\theta) + b_{nm}\cos(n\theta)\big)\cos(\beta_m z^*),$$

$$\tag{4.28}$$

where $\beta_m = \pi(2m-1)$. A_{nm} contains the solution of equation (4.27) and depends on the boundary condition from equation (4.9) which is prescribed at the outer bearing radius. Both solutions for A_{nm} are given in appendix C. Although for static problems the terms containing the sines would suffice to describe the pressure solution [31, 52], the cosine terms must be retained in order to solve dynamical problems. Note the mutual orthogonality of the solution components, which eventually leads to a particular sparse system matrix. It should be mentioned that these harmonic ansatz functions also lend them self very well to describe typical pressure solutions of the Reynolds equation. The derivative with respect to r^*, which is needed in the Reynolds equation is given by

$$\left.\frac{\partial \widehat{p}_{\mathrm{h,a}}^*(\theta, z^*, r^*)}{\partial r^*}\right|_{r^*=1} = \sum_{n,m} B_{nm}(r^* = 1)\big(a_{nm}\sin(n\theta) + b_{nm}\cos(n\theta)\big)\cos(\beta_m z^*), \tag{4.29}$$

where B_{nm} is given in appendix C for both types of boundary conditions prescribed at the outer bearing radius. Using the Reynolds equation (4.5) as a residual function for Galerkin's method produces the following weighted residual:

$$\int_{-1/2}^{+1/2}\int_{0}^{2\pi}\left\{\frac{\partial}{\partial\theta}\left(h^3\frac{\partial p^*_{h,a}}{\partial\theta}\right)+\kappa^2\frac{\partial}{\partial z^*}\left(h^3\frac{\partial p^*_{h,a}}{\partial z^*}\right)+12\Psi\left[\frac{\partial \widehat{p}^*_{h,a}}{\partial r^*}\right]_{r^*=1}-\Omega^*\frac{\partial h}{\partial\theta}-2\frac{\partial h}{\partial\tau}\right\}$$

$$\times\,\varphi_k(\theta)\cos(\beta_j z^*)\mathrm{d}\theta\mathrm{d}z^* = 0,$$

(4.30)

where $\varphi_k(\theta)=\sin(k\theta),\ \cos(k\theta)$. Symbolic evaluation of this integral, followed by some cumbersome algebra and appropriate re-arrangement of the equations eventually yields a formulation of the form

$$\sum_{n=0}^{N}a_{nj}\mathcal{F}_{njk}=c_{jk}\quad\text{and}\quad\sum_{n=0}^{N}b_{nj}\mathcal{G}_{njk}=d_{jk}.$$

(4.31)

For $k=0,\ldots,N$ these equations form two systems of linear equations for each $j=1,\ldots,M$ with the coefficients a_{nj} and b_{nj} as unknowns. As will be shown later on, only a small number of basis functions is needed to obtain a good approximation.

When setting up the problem matrices $\mathcal{F}_{njk}$ and $\mathcal{G}_{njk}$ the analytical orthogonality of the basis functions in equation (4.28) is exploited in order to predict the sparsity pattern of the matrices in advance. The symmetric positive definite problem matrices only have nonzero elements on the main diagonal and the three adjacent diagonals below and above the main diagonal. This information can be used to significantly accelerate the solution algorithm since it is a priori known which elements vanish and non-necessary computations can be avoided.

The approach allows for symbolic calculation of the bearing forces where only the zero crossings of the pressure functions (delimiting the region of positive pressures) have to be calculated numerically. Due to the fact that the derivative of the pressure functions is known analytically this can be performed efficiently using Newton's method. Combined with the low dimension and sparsity of the linear system, this results in very fast and accurate results compared to numerical discretization methods introduced in the next section (the reader is referred to section 5.1 for a detailed comparison). Moreover, it allows for symbolic derivation of stiffness and damping coefficients without the need to carry out numerical differentiation (see section 4.3.1). By setting $\Psi=0$ and leaving out the term A_{nm} from equation (4.28) the problem simplifies to the solid journal bearing problem.

It also is possible to use the film thickness function from equation (4.2) to account for misalignment. This requires a modified ansatz function, which allows asymmetrical pressure profiles in axial direction:

$$\widehat{p}^*_{h,a}(\theta,z^*,r^*=1)=\sum_{n,m}A_{nm}(r^*=1)\big(a_{nm}\sin(n\theta)+b_{nm}\cos(n\theta)\big)\sin\big(m\pi(z^*+1/2)\big).$$

(4.32)

By employing this new ansatz function and film thickness function the weighted residual has to be adjusted accordingly. This eventually leads to a system of linear equations of size $2NM \times 2NM$

$$\begin{bmatrix} \mathcal{F}_{nmkj} & \mathcal{H}_{nmkj} \\ \mathcal{J}_{nmkj} & \mathcal{G}_{nmkj} \end{bmatrix} \begin{bmatrix} a_{nm} \\ b_{nm} \end{bmatrix} = \begin{bmatrix} c_{kj} \\ d_{kj} \end{bmatrix}. \tag{4.33}$$

Although the left hand side still is a symmetric positive definite matrix it now contains several additional non-zero diagonals that increase the needed effort to solve the system of linear equations accordingly. A brief description of the sparse matrix algorithm, which is used to solve this system is given section 6.3.1.

When the coefficients are known the components of the fluid film force and moment can be calculated by integration over the positive domain of the pressure profile. Because the boundary of the positive pressure now depends on the axial *and* circumferential coordinate a new strategy is needed to determine the boundaries and calculate the hydrodynamic bearing forces and moments[2]. The pressure range is divided into smaller domains with linear boundaries as is illustrated in figure 4.3. The advantage of this approach is that the resulting integrals still solve to closed-form expressions. The hydrodynamic force and moment components are now given by

$$F_{h,\varepsilon}^* = 3 \sum_i \int_{z_i^*}^{z_{i+1}^*} \int_{\xi_{1,i}(z^*)}^{\xi_{2,i}(z^*)} p_h^*(\theta,z^*) \cos\theta \; \mathrm{d}\theta \mathrm{d}z^*, \quad F_{h,\gamma}^* = 3 \sum_i \int_{z_i^*}^{z_{i+1}^*} \int_{\xi_{1,i}(z^*)}^{\xi_{2,i}(z^*)} p_h^*(\theta,z^*) \sin\theta \; \mathrm{d}\theta \mathrm{d}z^* \tag{4.34}$$

and

$$M_{h,\varepsilon}^* = -3 \sum_i \int_{z_i^*}^{z_{i+1}^*} \int_{\xi_{1,i}(z^*)}^{\xi_{2,i}(z^*)} p_h^*(\theta,z^*) \sin\theta \; z^* \, \mathrm{d}\theta \mathrm{d}z^*, \quad M_{h,\gamma}^* = +3 \sum_i \int_{z_i^*}^{z_{i+1}^*} \int_{\xi_{1,i}(z^*)}^{\xi_{2,i}(z^*)} p_h^*(\theta,z^*) \cos\theta \; z^* \, \mathrm{d}\theta \mathrm{d}z^*. \tag{4.35}$$

The next two sections cover two extensions of the Galerkin-based bearing model for plain journal bearings: the evaluation of stiffness and damping coefficients and the inclusion of flow factors proposed by PATIR and CHENG [124, 125]. In principle these extensions are also applicable to the plain journal bearing with misalignment. However, for reasons of simplicity the

---

[2]Actually, the boundaries of the positive pressure range always depend on the axial coordinate! This can be seen e.g. for a pressure profile calculated with the REYNOLDS cavitation condition. For such a profile the rupture boundary for example is slightly curved in axial direction. However, when applying the GÜMBEL condition these boundaries always are independent from the axial coordinate for aligned axes.

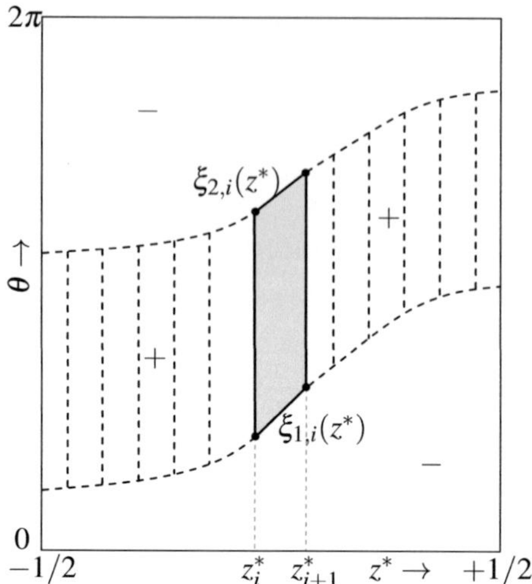

Figure 4.3: Schematic figure of the integration domain (positive pressure range) for the force and moment calculation in a misaligned plain journal bearing. The negative and positive part of the pressure range are denoted with $-$ and $+$.

following only deals with aligned bearings, i.e. $h = 1 + \varepsilon \cos\theta$ is used for the film thickness function.

4.3.1   Stiffness & damping coefficients

Linear elasticity and damping coefficients for small movements about the journal's equilibrium position often are very useful in machine design, for example to calculate resonance frequencies or to examine the stability of the equilibrium position. Usually, these coefficients are constructed by linearizing the force vector:

$$
\begin{bmatrix} F_{t,x}^* \\ F_{t,y}^* \end{bmatrix} = \begin{bmatrix} F_{t,x,0}^* \\ F_{t,y,0}^* \end{bmatrix} - \begin{bmatrix} k_{xx}^* & k_{xy}^* \\ k_{yx}^* & k_{yy}^* \end{bmatrix} \begin{bmatrix} \Delta\varepsilon_x \\ \Delta\varepsilon_y \end{bmatrix} - \begin{bmatrix} b_{xx}^* & b_{xy}^* \\ b_{yx}^* & b_{yy}^* \end{bmatrix} \begin{bmatrix} \Delta\varepsilon_x' \\ \Delta\varepsilon_y' \end{bmatrix},
\tag{4.36}
$$

in which the forces $F_{t,x,0}^*$ and $F_{t,y,0}^*$ are already known from a previous calculation performed for the given operating point. Stiffness and damping coefficients for unsteady states can also be used to construct the Jacobian matrix. Normally the Jacobian is calculated numerically using small perturbations to determine the derivatives, but if the Jacobian can be provided more accurately and in a more efficient manner the process of time integration can significantly gain performance and robustness. Here, this is shown by calculating the coefficients in combination with the Galerkin-based bearing model without using numerical differentiation. By assuming an equilibrium position, i.e. no translational velocity occurs, the coefficients simplify to their steady state version.

In general, however, the coefficients are calculated for a linearization point given by ε_x, ε_y, ε_x' and ε_y' for which the stiffness and damping coefficients are defined respectively by

$$k_{ij}^* = -\frac{\partial F_{\mathrm{t},i}^*}{\partial \varepsilon_j} \quad \text{and} \quad b_{ij}^* = -\frac{\partial F_{\mathrm{t},i}^*}{\partial \varepsilon_j'}, \tag{4.37}$$

where $i,j = x,y$. Substituting the transformed forces calculated with equation (4.10) and (4.11) results in

$$\frac{\partial F_{\mathrm{t},x}^*}{\partial \varepsilon_v^{(\prime)}} = +\frac{\partial F_{\mathrm{t},\varepsilon}^*}{\partial \varepsilon_v^{(\prime)}}\sin\gamma + \frac{\partial F_{\mathrm{t},\gamma}^*}{\partial \varepsilon_v^{(\prime)}}\cos\gamma + \left(F_{\mathrm{t},\varepsilon}\cos\gamma - F_{\mathrm{t},\gamma}\sin\gamma\right)\frac{\partial\gamma}{\partial\varepsilon_v^{(\prime)}}, \tag{4.38}$$

$$\frac{\partial F_{\mathrm{t},y}^*}{\partial \varepsilon_v^{(\prime)}} = -\frac{\partial F_{\mathrm{t},\varepsilon}^*}{\partial \varepsilon_v^{(\prime)}}\cos\gamma + \frac{\partial F_{\mathrm{t},\gamma}^*}{\partial \varepsilon_v^{(\prime)}}\sin\gamma + \left(F_{\mathrm{t},\varepsilon}\sin\gamma + F_{\mathrm{t},\gamma}\cos\gamma\right)\frac{\partial\gamma}{\partial\varepsilon_v^{(\prime)}}. \tag{4.39}$$

It is noted that the Galerkin-based model is only considered with the hydrodynamical part of the forces in this equation. After substitution of the pressure function from equation (4.28) the hydrodynamic force derivatives become

$$\frac{\partial F_{\mathrm{h},\varepsilon}^*}{\partial \varepsilon_v^{(\prime)}} = 3\cos\theta_2\frac{\partial\theta_2}{\partial\varepsilon_v^{(\prime)}}\int_{-1/2}^{+1/2}p_{\mathrm{h}}^*(\theta_2,z^*)\mathrm{d}z^* - 3\cos\theta_1\frac{\partial\theta_1}{\partial\varepsilon_v^{(\prime)}}\int_{-1/2}^{+1/2}p_{\mathrm{h}}^*(\theta_1,z^*)\mathrm{d}z^*$$

$$+3\int_{-1/2}^{+1/2}\int_{\theta_1}^{\theta_2}\left\{\sum_{n,m}A_{nm}(r^*=1)\left(\frac{\partial a_{nm}}{\partial\varepsilon_v^{(\prime)}}\sin(n\theta) + \frac{\partial b_{nm}}{\partial\varepsilon_v^{(\prime)}}\cos(n\theta)\right)\cos(\beta_m z^*)\right\}\cos\theta\,\mathrm{d}\theta\,\mathrm{d}z^*, \tag{4.40}$$

$$\frac{\partial F_{\mathrm{h},\gamma}^*}{\partial \varepsilon_v^{(\prime)}} = 3\sin\theta_2\frac{\partial\theta_2}{\partial\varepsilon_v^{(\prime)}}\int_{-1/2}^{+1/2}p_{\mathrm{h}}^*(\theta_2,z^*)\mathrm{d}z^* - 3\sin\theta_1\frac{\partial\theta_1}{\partial\varepsilon_v^{(\prime)}}\int_{-1/2}^{+1/2}p_{\mathrm{h}}^*(\theta_1,z^*)\mathrm{d}z^*$$

$$+3\int_{-1/2}^{+1/2}\int_{\theta_1}^{\theta_2}\left\{\sum_{n,m}A_{nm}(r^*=1)\left(\frac{\partial a_{nm}}{\partial\varepsilon_v^{(\prime)}}\sin(n\theta) + \frac{\partial b_{nm}}{\partial\varepsilon_v^{(\prime)}}\cos(n\theta)\right)\cos(\beta_m z^*)\right\}\sin\theta\,\mathrm{d}\theta\,\mathrm{d}z^*, \tag{4.41}$$

where $v = x,y$. Here $\partial a_{nm}/\partial\varepsilon_v^{(\prime)}$ and $\partial b_{nm}/\partial\varepsilon_v^{(\prime)}$ are the new unknown coefficients that need to be calculated. One way to obtain these is to differentiate the system of linear equations in equation (4.31) with respect to $\varepsilon_v^{(\prime)}$. This results in

$$\sum_{n=0}^{N}\mathcal{F}_{njk}\frac{\partial a_{jk}}{\partial \varepsilon_v^{(\prime)}} = \frac{\partial c_{jk}}{\partial \varepsilon_v^{(\prime)}} - \sum_{n=0}^{N}\frac{\partial \mathcal{F}_{njk}}{\partial \varepsilon_v^{(\prime)}}a_{jk} \quad \text{and} \quad \sum_{n=0}^{N}\mathcal{G}_{njk}\frac{\partial b_{jk}}{\partial \varepsilon_v^{(\prime)}} = \frac{\partial d_{jk}}{\partial \varepsilon_v^{(\prime)}} - \sum_{n=0}^{N}\frac{\partial \mathcal{G}_{njk}}{\partial \varepsilon_v^{(\prime)}}b_{jk}, \quad (4.42)$$

where the coefficients a_{jk} and b_{jk} for the current operating point follow from the original Galerkin-based bearing model (see equation (4.31)). Again, for $k = 0,\ldots,N$ these form two systems of linear equations for each $j = 1,\ldots,M$. In solving these systems of equations the unknown coefficients are obtained. Comparisons of these stiffness and damping coefficients with analytical solutions and other results for porous journal bearings are shown in chapter 5.

4.3.2   Surface roughness with flow factors

The influence of rough surfaces on the hydrodynamical pressure is introduced into the model by using the averaged Reynolds equation [125]

$$\frac{\partial}{\partial \theta}\left(\phi_\theta^f h^3 \frac{\partial p_h^*}{\partial \theta}\right) + \kappa^2 \frac{\partial}{\partial z^*}\left(\phi_z^f h^3 \frac{\partial p_h^*}{\partial z^*}\right) + 12\Psi \left[\frac{\partial \widehat{p}_h^*}{\partial r^*}\right]_{r^*=1} = \frac{\partial h}{\partial \theta} + 2\frac{\partial h}{\partial \tau} + \frac{\sigma_h}{c}\frac{\partial \phi_s^f}{\partial \theta}, \quad (4.43)$$

where ϕ_θ^f and ϕ_z^f are respectively the pressure flow factors in circumferential and axial direction, ϕ_s^f is the shear flow factor and σ_h is the standard deviation of the combined roughness of both contact surfaces. It is assumed that the surface roughness distribution is Gaussian. The flow factors are functions of the film thickness and their influence vanishes for smooth surfaces and/or large film thickness, i.e.

$$\phi_\theta^f, \phi_z^f \to 1 \quad \text{and} \quad \phi_s^f \to 0 \quad \text{as} \quad H/\sigma_h \to \infty.$$

A flow factor can be determined as a function of film thickness with small scale hydrodynamic simulations as was shown in section 2.1.2. This work is concerned with the implications of the flow factors on journal bearing characteristics. Therefore, only an implementation (and no numerical calculation) of flow factors into the previously shown Galerkin-based model is presented here. The flow factors calculated from [124, 125] are chosen as test cases and the reader is referred to appendix D for their complete expressions. To make an implementation possible with the Galerkin-based model the flow factors need to be approximated using simplified analytical functions. For the pressure flow factors

$$\phi_{\theta,z}^{f} \cong \sum_{i=0}^{6} C_{i}^{\theta,z} h^{i-3} \tag{4.44}$$

is proposed in which the coefficients $C_{i}^{\theta,z}$ are to be determined through curve fitting with the original flow factor form. With this approximation orthogonality in the weighted residual integral is preserved. For the shear flow factor the following function is proposed

$$\Phi_{s} \cong C_{0}^{s} \frac{h}{C_{1}^{s} + (h + C_{2}^{s})^{C_{3}^{s}}}, \tag{4.45}$$

where again the coefficients C_{i}^{s} are to be determined through curve fitting. Note that the shear flow factor ϕ_{s}^{f} depends on Φ_{s} as described in appendix D.

The weighted residual integral has to be modified accordingly to equation (4.43), the remaining steps in solving the problem and calculating bearing characteristics remain unchanged.

4.4 Approximations with numerical discretization methods

By using numerical methods it becomes possible to employ the advanced cavitation boundary conditions which were briefly discussed in chapter 2. Using numerical iteration schemes the moving boundary problem can now be solved, which occurs when cavitation using the REYNOLDS or JFO boundary condition is included. Application of the JFO condition is omitted here, but a short description of the REYNOLDS condition implemented for a numerical iteration scheme is given here. Within this work two numerical methods were used to verify the proposed bearing model. A numerical iteration scheme using a FD discretization based on [158] was developed and will be described here in more detail. Its correctness for porous bearings was verified by comparing the results to those of the commercial FEM software COMSOL[3] of which a detailed description is omitted here for reasons of brevity.

---

[3]© COMSOL Multiphysics GmbH

The FD discretization of equation (4.5) for the left hand side involves the terms [153]

$$\frac{\partial}{\partial\theta}\left(h^3\frac{\partial p_h^*}{\partial\theta}\right) = \frac{\zeta_{i-1/2,j}p_{i-1,j}^* - (\zeta_{i-1/2,j} + \zeta_{i+1/2,j})p_{i,j}^* + \zeta_{i+1/2,j}p_{i+1,j}^*}{h_\theta^2}, \quad (4.46)$$

$$12\Psi\left[\frac{\partial\widehat{p}_h^*}{\partial r^*}\right]_{r^*=1} = 12\Psi\frac{\widehat{p}_{i,j,k=2}^* - \widehat{p}_{i,j,k=1}^*}{h_r} \quad (4.47)$$

where $\zeta_{i\pm1/2,j} = (\zeta_{i,j} + \zeta_{i\pm1,j})/2$, $\zeta_{i,j} = h_{i,j}^3$ and h_θ and h_r indicate the grid size in circumferential and radial direction. The discretization of the first term is completely analogous to the second term on the left hand side. For the used film thickness function the right hand side of equation (4.5) is known analytically and derivatives may be calculated in an explicit form, so that it does not need to be approximated with finite differences.

When including the REYNOLDS cavitation condition the Reynolds equation becomes nonlinear and an iterative scheme has to be used to solve the resulting moving boundary problem. Because the rupture and reformation points of the fluid film are a priori unknown, implementation of cavitation boundaries into a numerical scheme has to succeed alongside these iterations. Venner et al. [153] implemented this in a SOR or Gauss-Seidel iteration scheme, where every negative pressure is immediately set to zero during the iteration procedure. When a converged solution is found the Reynolds condition is automatically satisfied. This algorithm is mostly applied to FD methods which are solved through iteration schemes.

A further algorithm which delivers the same result is Murty's algorithm or the so-called penalty method, which is based on [109]. Nodes inside the cavitation area with negative pressure are compensated in a succeeding iteration using a so-called penalty factor. Repeating this process also results in a converged solution which is identical to the solution of the previously described method. This algorithm mostly is used in combination with FEM [86] and also is used in COM-SOL to simulate the REYNOLDS condition.

For porous journal bearings Darcy's law has to be solved additionally. The FD discretization of equation (4.6) involves the terms

$$\frac{\partial^2\widehat{p}_h^*}{\partial r^{*2}} = \frac{\widehat{p}_{i,j,k-1}^* - 2\widehat{p}_{i,j,k}^* + \widehat{p}_{i,j,k+1}^*}{h_r^2} \quad (4.48)$$

$$\frac{\partial\widehat{p}_h^*}{\partial r^*} = \frac{\widehat{p}_{i,j,k-1}^* - \widehat{p}_{i,j,k}^*}{h_r}. \quad (4.49)$$

The other two terms in equation (4.6) are discretized analogously to those in equation (4.48). The effect of cavitation is ignored for the porous medium and it therefore is not necessary to employ iteration schemes to solve this equation. Because Darcy's law has no terms depending

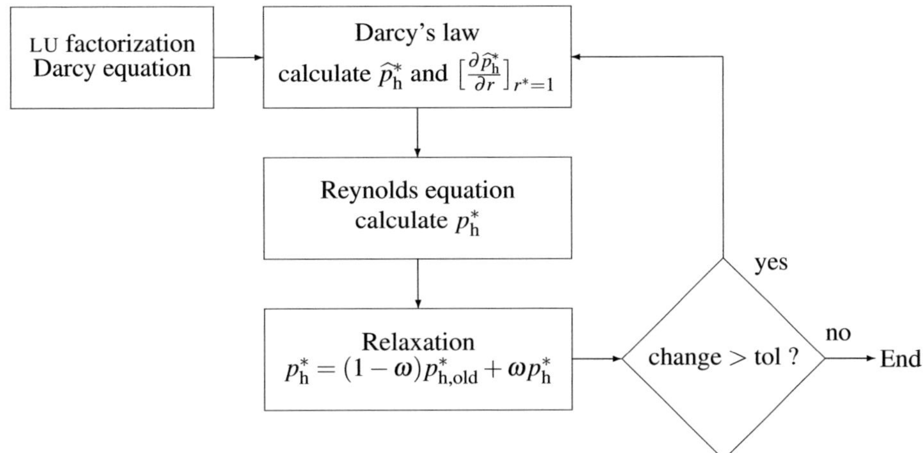

Figure 4.4: Iteration scheme for the FD method for porous bearings. ω is a relaxation parameter.

on one or more of the variables (such as the film thickness function $h(\theta, z^*)$ which appears in the Reynolds equation) this equation can be factorized prior to the iteration procedures, that are necessary e.g. for cavitation algorithms or time integration. In doing so, the factorization procedure of the left hand side of the system of linear equations only has to be performed one single time at the beginning of the simulation procedure. Normally, this procedure accounts for a significant part of the calculation time. Another advantage of this method is, that by reorganizing the system matrix the number of non-zero entries can also be significantly reduced. This technique is employed for the FD method and can also be carried out prior to the actual calculations.

Finally, to obtain a converged solution for the system of coupled differential equations an additional iteration scheme is required. First, a converged solution of the Reynolds equation is obtained without the influence of the porous matrix (this solution corresponds to $p^*_{\text{h,old}}$ in figure 4.4 during the first iteration). This pressure field can then be used to solve the governing equation for the porous medium. It is now possible to solve the Reynolds equation including the exchange term for flow from and to the porous medium. With the resulting pressure the process returns to solving the Darcy equation. This procedure is repeated until a converged solution has been found. To avoid numerical oscillations it is important to limit the influence per iteration of the porous matrix into the pressure solution of the Reynolds equation. To this end a relaxation parameter is introduced, which has to be chosen such that convergence is guaranteed, but the calculation time still remains acceptable. The iteration process is illustrated in figure 4.4.

Results for solid and porous journal bearings using this approach are shown in the following chapter.

Chapter 5

Plain journal bearing characteristics

Using the models that were introduced in the previous chapter the steady state characteristics of a single plain journal bearing will now be studied.

To assure a proper assessment of the physical characteristics the proposed method's numerical performance and accuracy will first be studied to establish, amongst other things, the necessary number of ansatz functions. This may change depending on whether a porous or solid bearing is considered and on the amount of misalignment. Furthermore, a first indication of the method's validity is given by verifying it with alternative models and cavitation algorithms that were introduced in the previous chapter. The assessment is concluded with an experimental validation.

To study the characteristics of a single plain journal bearing, the solid journal bearing problem is solved using Galerkin's method. Pressure profiles as well as steady state properties are compared to known analytical solutions and full numerical solutions. Subsequently, the method is applied to a porous journal bearing after which its flexibility is demonstrated by applying it to the generalized formulation of the Reynolds equation (equation (4.43)) to account for the influence of rough surfaces [151]. Followed by this the added influence of asperity contacts is looked into and finally the effects of misalignment on the hydrodynamical load and moment capacity are studied.

5.1 Numerical performance and accuracy

It is important to choose a sufficient number of ansatz functions for the Galerkin-based model to obtain sufficient accuracy. In view of performance and the choice of plain journal bearing

Method	Unknowns			Load capacity (F_{h}^*)	Δ [%]	Calc. time [s]
	n_x	n_y	Total			
FD	24	4	96	9.8674	13.61	0.004
FD	48	8	384	11.0159	3.55	0.020
FD	96	16	1536	11.3190	0.90	0.087
FD	192	32	6144	11.3961	0.22	0.363
FD	384	64	24576	11.4155	0.05	2.056
FD	768	128	98304	11.4203	0.01	73.165
FD	1536	256	393216	11.4216	-	>100
	N	M	Total			
Galerkin	6	4	48	11.1152	2.68	0.002
Galerkin	10	5	100	11.4147	0.06	0.003
Galerkin	15	10	300	11.4212	0.004	0.003
Galerkin	25	15	750	11.4216	-	0.004

Table 5.1: Comparison of the load capacity calculated with the FD scheme and the Galerkin-based bearing model for different numbers of unknowns. Here n_x and n_y represent the number of grid points in resp. circumferential and axial direction for the FD scheme. Δ denotes the relative deviation from the converged value [151].

type the acceptable number of ansatz functions may vary and therefore it also is useful to study their influence on performance and accuracy.

Aligned bearings

In order to assess the performance of the proposed bearing model it is compared to the numerical FD scheme which was introduced in the previous chapter. This is done for an arbitrary dynamical case ($\varepsilon_x = 0.5$, $\varepsilon_x' = 0.9$, $\varepsilon_y = -0.5$ and $\varepsilon_y' = 0.3$) for the solid journal bearing problem with the GÜMBEL cavitation condition. For other dynamical cases similar results were obtained. The number of unknowns as well as the convergence of the solution – characterized by the dimensionless hydrodynamic load capacity F_{h}^* – and the calculation time are compared. Both approaches yield systems of linear equations which are solved on the same computer using the same direct sparse solver. The measurements of calculation time only account for the calculations that need to be performed within a single time iteration, i.e. any preparations that only have to be performed once prior to a simulation are not taken into account. The results are arranged in table 5.1 [151].

From these results it becomes clear that to reach a converged solution the amount of unknowns for the FD scheme is orders of magnitude larger than for the Galerkin-based model. Aside from longer calculation times the memory demands should now also be regarded. The advantages of the model using Galerkin's method with global ansatz functions over a FD scheme for the plain

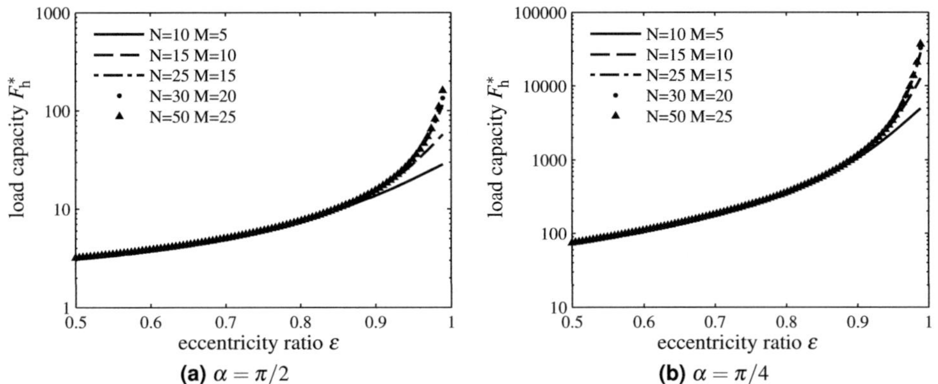

Figure 5.1: Convergence of the hydrodynamic load capacity for Galerkin's method in a solid journal bearing with $L/D = 1$ for different α.

journal bearing become even more apparent for the porous journal bearing problem. Compared to the plain journal bearing the number of unknowns in a porous journal bearing for the FD scheme will increase significantly. This has to do with the fact that the pressure field in the porous medium needs to be discretized additionally. However, for the proposed model using Galerkin's method the added law of Darcy can be solved analytically and included in the basis function a priori, which leads to practically no additional numerical costs. When the eccentricity ratio is raised an increasing number of ansatz functions is needed to obtain a usable solution, especially when the eccentricity ratio approaches one. This in turn also depends strongly on

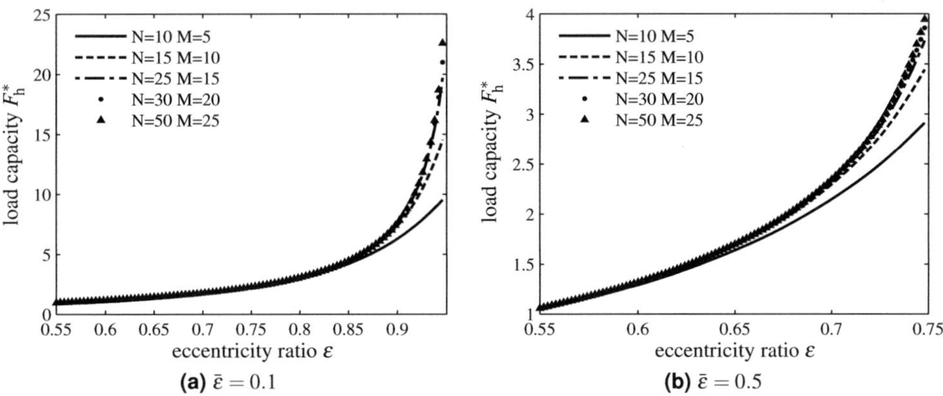

Figure 5.2: Dimensionless hydrodynamic load capacity versus eccentricity ratio for different degrees of misalignment and number of ansatz functions in a solid journal bearing with $L/D = 1$ and $\psi = 0$.

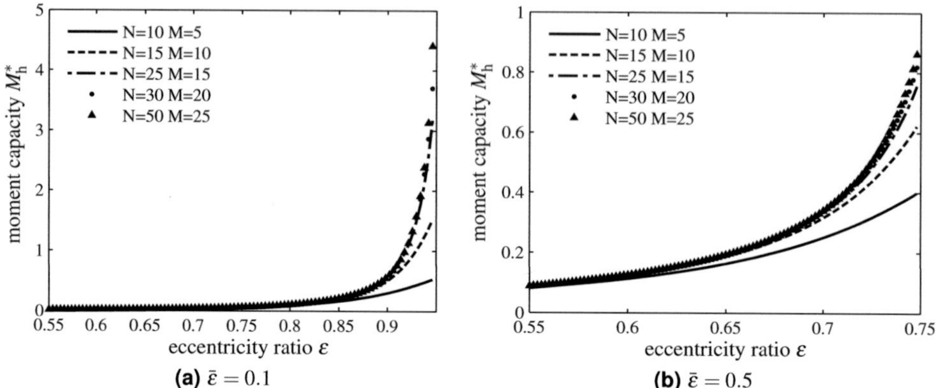

(a) $\bar{\varepsilon} = 0.1$ **(b)** $\bar{\varepsilon} = 0.5$

Figure 5.3: Dimensionless hydrodynamic moment capacity versus eccentricity ratio for different degrees of misalignment and number of ansatz functions in a solid journal bearing with $L/D = 1$ and $\psi = 0$.

other parameters such as the permeability of the porous medium and the tilting angle of the journal axis compared to the bearing axis.

To establish the necessary number of ansatz functions the convergence of the total load capacity versus eccentricity ratios is observed for different dynamical situations (indicated by the angle α). These are shown in figure 5.1. Even for a relatively high eccentricity ratio of $\varepsilon = 0.95$ a small number of ansatz function suffices to obtain a relatively good approximation. At very large eccentricity ratios, however, problems may occur. This has to do with the fact that for such parameters a very narrow pressure spike is situated near the minimum film thickness, which requires many ansatz functions. Fortunately, for porous journal bearings these pressure spikes are less pronounced, especially at very high eccentricity ratios. Here, the pressure build-up is counterbalanced by seepage flow into the porous medium, which prevents large pressure spikes. The occurrence of pressure spikes will be illustrated for a misaligned bearing in the following.

Misaligned bearings

The issues occurring at very high eccentricity ratios for the model using Galerkin's method in solid journal bearings become even more apparent for misaligned journal bearings. This is illustrated by observing the total hydrodynamical load and moment capacity versus the eccentricity ratio for different rates of misalignment (indicated by $\bar{\varepsilon}$) and a varying number of ansatz functions. These are shown in figure 5.2 for the load capacity F_{h}^* and figure 5.3 for the mo-

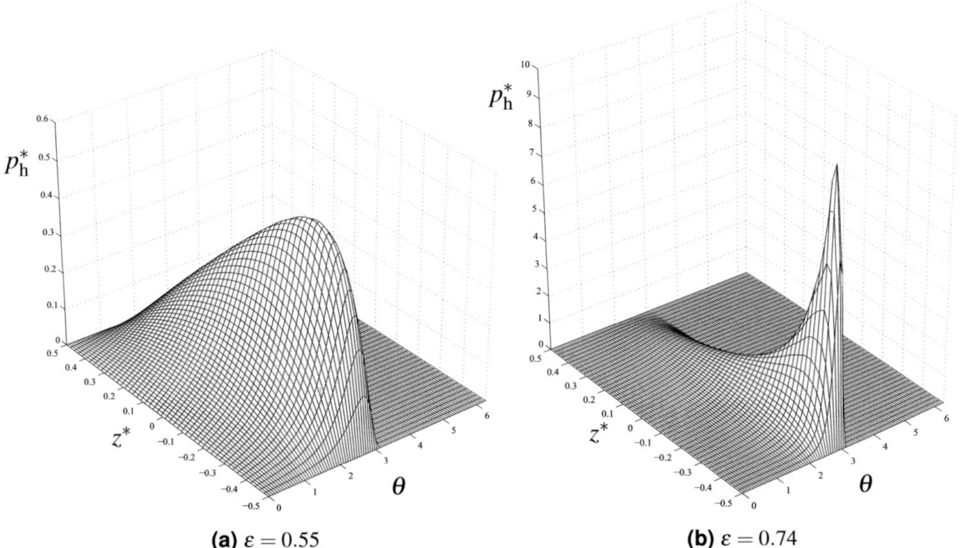

(a) $\varepsilon = 0.55$
(b) $\varepsilon = 0.74$

Figure 5.4: Pressure profiles for a misaligned solid journal bearing with $L/D = 1$, $\bar{\varepsilon} = 0.5$ and $\psi = 0$. The number of ansatz functions was set to $N = 50$ and $M = 25$.

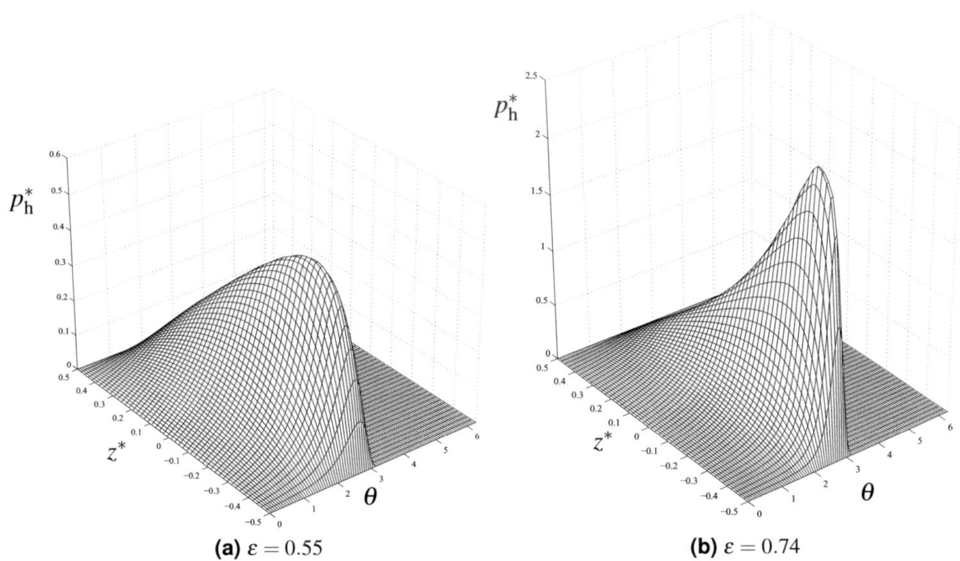

(a) $\varepsilon = 0.55$
(b) $\varepsilon = 0.74$

Figure 5.5: Pressure profiles for a misaligned porous journal bearing with $L/D = 1$, $\bar{\varepsilon} = 0.5$, $\psi = 0$, $\Psi = 0.001$ and $R_o/R_i = 2$. The number of ansatz functions was set to $N = 50$ and $M = 25$.

ment capacity M_h^*. It is seen that for the chosen number of ansatz functions these converge to a finite value when the local eccentricity ratio goes to 1. However, it has been shown that also for these situations the load and moment capacity approach infinity when the local eccentricity ratio goes to 1 [15]. Whereas for aligned solid journal bearings the pressure spikes were more or less stretched out over the complete axial length of the bearing, here, the pressure spike is localized at one axial end. Pressure profiles for a solid journal bearing with $\bar{\varepsilon} = 0.5$ are shown in figure 5.4 to illustrate this. Particularly many ansatz functions are required to construct such pressure profiles. For misaligned porous journal bearings these problems are not an issue, because here the formation of localized pressure spikes is less pronounced for the same reasons depicted for aligned porous journal bearings. To illustrate this pressure profiles for a porous journal bearing with otherwise identical parameters as in figure 5.4 are shown in figure 5.5.

5.2 Model verification and validation

A first indication of the validity of the proposed bearing model is given by comparing it to numerical methods with different cavitation conditions. Impedance maps are used to compare the different cavitation algorithms that were introduced in the previous chapter, see figure 5.6. The map with the GÜMBEL cavitation condition has been generated with the proposed bearing model using a Galerkin discretization. The other map has to be constructed using a numerical

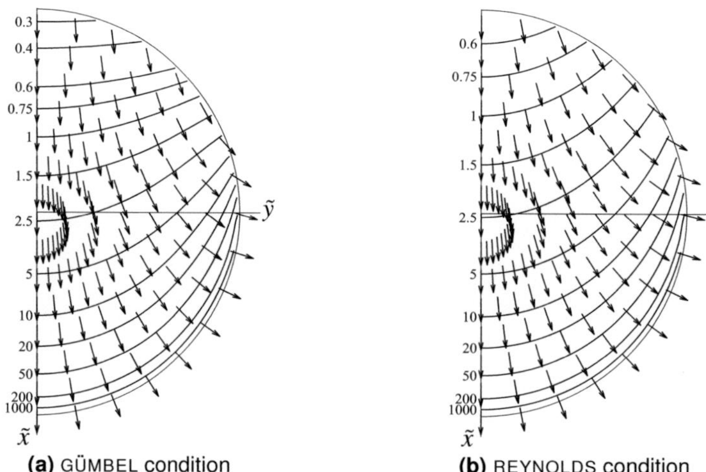

(a) GÜMBEL condition   (b) REYNOLDS condition

Figure 5.6: Impedance maps for a solid journal bearing with $L/D = 1$ for two different cavitation algorithms.

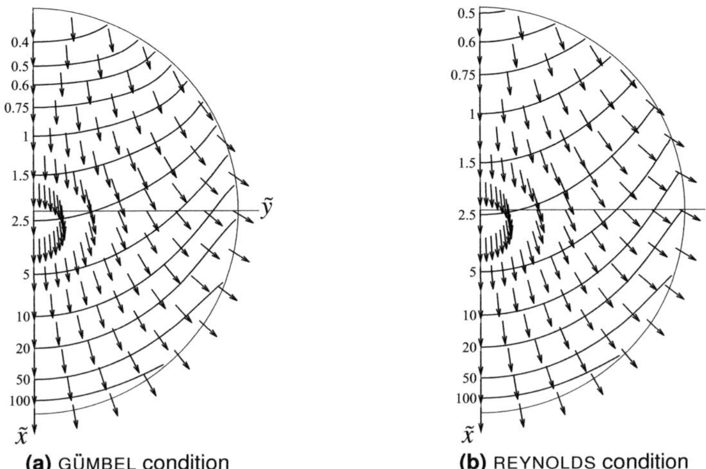

(a) GÜMBEL condition   (b) REYNOLDS condition

Figure 5.7: Impedance maps for a porous journal bearing with $L/D = 1$, $\Psi = 0.001$ and $R_o/R_i = 2$ for two different cavitation algorithms.

discretization method involving an iteration scheme to be able to implement the REYNOLDS cavitation condition. Here the FD method has been used which was introduced in section 4.4. Although the difference between the pressure curves for the GÜMBEL and REYNOLDS conditions is apparent (see figure 2.2) this is not the case for the impedance maps. A deviation is observed in the upper middle part of the map where the impedance of the GÜMBEL condition is somewhat lower than for the REYNOLDS condition. However, on the whole the GÜMBEL cavitation condition approximates the solution with the REYNOLDS condition very well. Since the proposed bearing model produces the same result as a FD scheme with the GÜMBEL cavitation condition it has been omitted here. The reader is referred to [152] for this impedance map. For the porous journal bearing a similar agreement is observed as for the solid journal, the difference seen in the upper middle part is now even smaller.

To validate the proposed model several experiments were performed. Two types for porous journal bearings were tested in a special experimental setup for porous journal bearings. The reader is referred to [150] for a detailed overview of this setup, its properties and capabilities. Due to the fact that the shown experimental results were part of a larger study with partly different goals only some results have been used to validate the proposed bearing model.

The experimental setup consists of a shaft which is supported by four bearings. Torque can be applied separately to each bearing. Using this setup the total friction moment of these four bearings can be measured as well as the temperature at three positions in the experimental setup

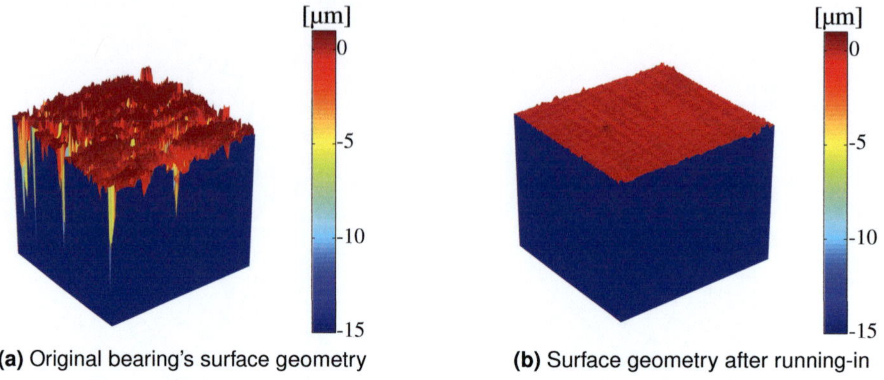

(a) Original bearing's surface geometry   **(b)** Surface geometry after running-in

Figure 5.8: Surface geometries of a small patch of the bearing's surface.

located on the bearings. Additionally, the journal's position with respect to the bearings is measured along one axis of the coordinate system.

To be able to make a comparison the temperature measurements are used to determine the viscosity of the lubricant used in the simulation. The permeability K was determined by measuring the amount of air flow through the porous medium in a separate experiment. The regime of mixed lubrication is modeled with the GREENWOOD & WILLIAMSON asperity contact model described in section 2.1.2. In order to use this model several surface roughness parameters are required. The surface geometry of a small patch of the bearing surface before and after running-in was obtained using white-light interferometry. Figure 5.8 shows the measured surface geometries. After analysis the asperity peak-height's density $\hat{\eta}$, standard deviation σ_e and mean asperity curvature radius $\hat{\beta}$ were obtained. The composite elastic modulus can be obtained from literature and the Coulomb friction coefficient parameter is assumed at $\mu_c = 0.15$. This is an average from a series of values for lubricated steel to steel contacts from literature [11]. Since only the surface geometry of the bearing was measured the journal's surface is assumed smooth and the asperity contact model from equation (2.13) for a smooth to rough contact is used.

Measurements were performed for sintered bronze and iron journal bearings. Here, only the results for bronze porous journal bearings are presented for low and high loads. Results for the sintered iron bearings had to be discarded due to prolonged running-in behavior, which resulted in almost no hydrodynamic lubrication for almost all sintered iron bearing cases. The reasons for this will be discussed later on. Table 5.2 summarizes the experimental data which have been used for the parametrization of the simulation (with the exception of torque). To compare the experiments to the simulation results the friction coefficient as a function of the journal's surface velocity is used. This coefficient was determined for selected cycles from a total of 150 consecutive cycles for each measurement series. A single cycle consists of a linear increase of surface velocity to a value of 2.5 m/s followed by a linear decrease to zero velocity. Figure 5.9 shows an example of the measured friction coefficient against the surface velocity for 150 consecutive cycles. Here, only the friction coefficients are shown for increasing surface velocity. It should also be noted that the values in the mixed lubrication regime, close to vanishing surface velocity, are left out for all experimental results of the friction coefficient, due to the fact that these were deemed implausible when comparing them to other measurements of porous bearings, e.g. in [18]. Depending on the type of bearing and the applied load a certain running-in is observed when comparing the friction coefficient of subsequent cycles. On the one hand, this phenomenon smoothes the surface, such that it is more likely for a fluid film to form. However, on the other hand does it also diminish the permeability of the porous bearing through pore closure. It is therefore important to compare the proposed model to measurements for which the bearing's surface has been run-in sufficiently and a fluid film could form. Thus, for all comparisons the friction coefficient was taken after a certain time of running-in and the roughness parameters corresponding to figure 5.8b were used to calculate the (friction) forces resulting from asperity contacts.

The comparisons of the friction coefficient for low and high loads are shown in figures 5.10 and 5.11 respectively. Two different models were used for the comparison with the experimental results: a hydrodynamical model for solid journal bearings using the REYNOLDS cavitation condition (also used to calculate the impedance maps in figures 5.6 and 5.7) and the proposed bearing model based on the Galerkin discretization. The results for the proposed bearing model were calculated in ADAMS using a subroutine which will be briefly described in section 6.3.1 in the next chapter. The friction calculation in this last model is given by equation (4.16), where for the porous journal bearing model the friction calculation is modified as explained in section 4.1.2. All of the used models only account for aligned plain journal bearings. It has been observed that the proposed bearing model for misaligned porous journal bearings cannot cope with the considered applied torques under hydrodynamic conditions and even not when elastic asperity contacts are included. This can be explained partly due to the fact that the fluid film geometry is derived for a perfect cylindrical journal bearing, which results in a relative

load [N]	torque [N cm]	clearance [µm]	cycle	viscosity [Pa s]
250	90	26	150	0.020847
250	0	13	102	0.021281
50	30	26	150	0.03129
50	90	6	150	0.022877

Table 5.2: Overview of the measurement parameters for bronze porous journal bearings.

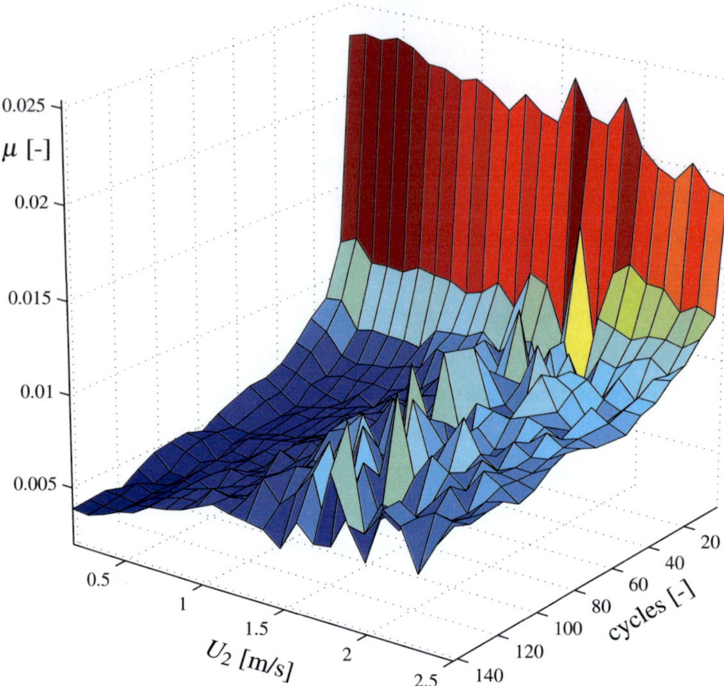

Figure 5.9: The friction coefficient μ vs. surface velocity U_2 for different subsequent cycles. Please note that not all cycles are shown here, because only a selection of all performed 150 cycles was measured.

small contact area with elastic asperity contacts for misaligned configurations. However, several other effects also can come into play for such high torques, which shall be addressed in the discussion at the end of this section.

As is seen in figure 5.10 a good approximation in the hydrodynamic range is made for low loads using the porous journal bearing model up to about 1.2 m/s. Here, a deviation in the friction coefficients is observed, which is suspected to originate from a resonance frequency of the experimental setup for this rotor speed. This effect is also observed in figure 5.9. For the proposed bearing model the onset of mixed lubrication already starts when the experimental results still seem to be in the hydrodynamic regime, which could have several causes. The surface roughness parameters were obtained for only one surface sample and it is very likely that for other journal bearing geometries and loading conditions the roughness properties differ. Even for low loads running-in could have diminished the bearing's permeability and this in turn also delays the onset of mixed lubrication. It is noted that the friction coefficients calculated with the proposed bearing model, denoted in the figures as ADAMS, converge to the prescribed

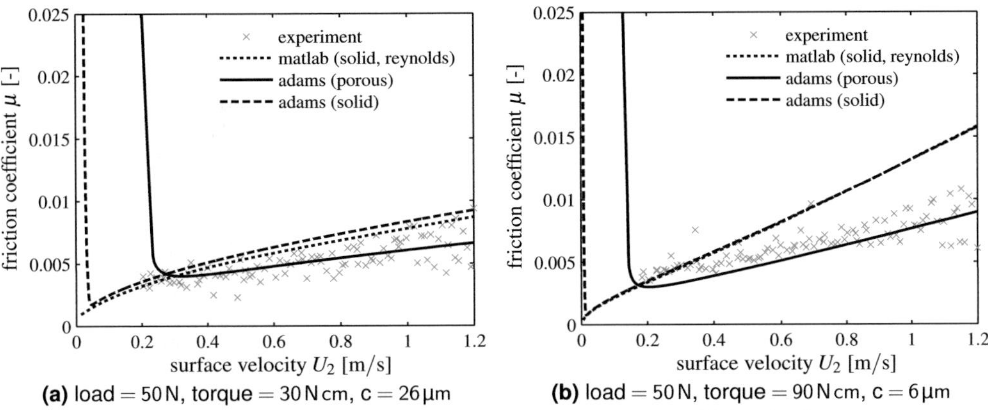

(a) load $= 50\,$N, torque $= 30\,$N cm, c $= 26\,\mu$m **(b)** load $= 50\,$N, torque $= 90\,$N cm, c $= 6\,\mu$m

Figure 5.10: Comparisons of the bronze bearing results with low loads.

value of μ_c for vanishing surface velocity. For reasons of clarity the figures only focus on the hydrodynamic range. This also is the case for the experiments with high loads.

These experiments can be found in figure 5.11 and show a very different behavior. Here, the porous journal bearing model predicts asperity contacts for a large surface velocity range, which is clearly not the case. For these high loads the running-in has a significant influence on the surface roughness but also on the permeability of the interface, which probably decreased significantly due to local pore closure. The latter explains the good agreement between the measurements and the solid journal bearing models. Also note the small difference in the hydrodynamical regime between the more accurate REYNOLDS model and the proposed bearing model, which is consistent with the differences observed between the impedance maps in figure 5.6. Although it seems that a porous journal bearing performs better with respect to load capacity when the majority of the pores have closed, the reader should keep in mind that this is not necessarily a good thing. Extensive pore closure also obstructs lubricant flow between the fluid film and the porous medium, which might shorten the porous bearing's life time significantly due to a shortage of lubricant! The optimal degree of pore closure will have to be a compromise between maximizing the load capacity and obtaining optimal lubrication conditions.

Discussion

The shown friction coefficients reflect two extreme loading cases of porous journal bearings. To be able to predict the friction coefficient for loading situations that lie in between these two extremes it is necessary to properly estimate the roughness parameters as well as the permeabil-

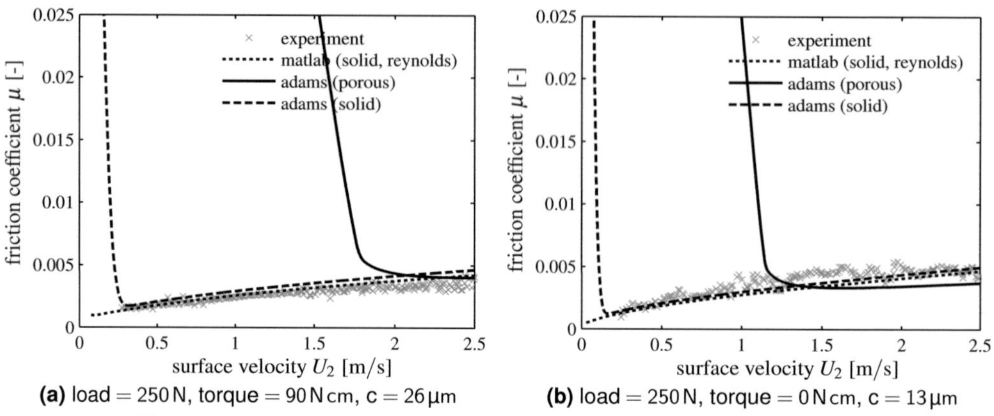

(a) load $= 250\,\mathrm{N}$, torque $= 90\,\mathrm{N\,cm}$, $c = 26\,\mu\mathrm{m}$ **(b)** load $= 250\,\mathrm{N}$, torque $= 0\,\mathrm{N\,cm}$, $c = 13\,\mu\mathrm{m}$

Figure 5.11: Comparisons of the bronze bearing results with high loads.

ity. These parameters can be either obtained through experiments or by numerical simulation of the run-in process for example by employing the BOUSSINESQ-CERUTTI theory [12]. The former was not part of the current experimental investigation and the latter is beyond the scope of this work. However, the current study has demonstrated that it is necessary to have proper estimates of the roughness and material parameters to accurately describe the friction coefficient of a porous journal bearing.

Although the current models are able to predict the friction coefficients in the hydrodynamic regime for sintered bronze bearings relatively well, the same agreement could not be established for the sintered iron bearings. Due to the fact that the yield stress of the chosen sintered iron is higher than of sintered bronze their running-in behavior differs significantly. Sintered iron bearings will therefore generally operate longer under mixed lubrication conditions. This also has been observed in experiments where many runs did not reach a stage where the bearings could be operated under hydrodynamical conditions. Since the current models are not able to describe the running-in mechanisms and their influence on the permeability and surface roughness, the comparisons with these measurements have been omitted.

Another issue is the influence of misalignment, which was not taken into account for the simulation results in the current comparisons. Apart from the issues that occur at very high eccentricity ratio with misalignment for the proposed bearing model at the axial ends (see the previous section), there also are other effects which should be considered in this regime. The local pressure concentration for misaligned bearings, for example, results in macroscopic elastic deformation and increases the total area with carrying pressure and asperity contacts. The predefined bearing shape at the axial ends (which differs from the perfect geometry assumed in section 4.1) also contributes to this effect. Moreover, localized pore-closure and special additives in the lubricant (graphite particles, that influence the lubricant behavior at very small film thicknesses) are now important factors that should be taken into account for these operating conditions. The current model cannot account for these effects, because they would severely increase the calculation time and make it therefore unpractical for applications such as rotor-dynamics analysis or multibody system simulation.

For an estimation of the influence of elastic deformation in aligned porous journal bearings the reader is referred to the frame below.

Note on the influence of macroscopic elastic deformation

Several numerical simulations were performed to assess the influence of the macroscopic deformation of the bush on the load capacity of porous journal bearings. The influence is constituted through the interaction between the fluid film pressure governed by the Reynolds equation and Darcy's law and the bearing's elastic deformation governed by Hooke's law, which is called elastohydrodynamic lubrication. The commercial software COMSOL has been used to model this problem. Since this software bases on the FEM a flexible bush is easily included into the model. For the fluid film a modified thickness function is used: $h = 1 + \varepsilon \cos \theta + u$. Here, $u = U/c$, where U refers to the deformation of the bearing's surface. The bearing's deformation is governed by Hooke's law which has the fluid pressure as boundary condition. For reasons of simplicity

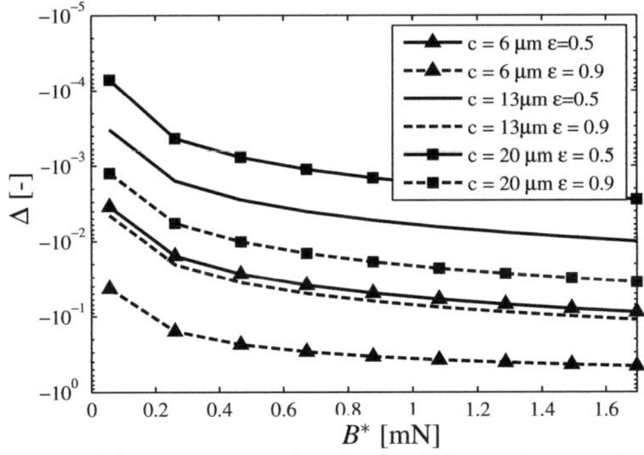

Figure 5.12: Influence of the macroscopic elastic deformation on the load capacity of a porous journal bearing.

Hooke's law is formulated in non-dimensionless form:

$$\sigma_{ij}^{\mathrm{el}} = \mathscr{L}_{ijkl}\,\varepsilon_{kl}^{\mathrm{el}}, \tag{5.1}$$

where $\sigma_{ij}^{\mathrm{el}}$ refers to the stress tensor, $\mathscr{L}_{ijkl}$ is the elasticity tensor and $\varepsilon_{kl}^{\mathrm{el}}$ the strain tensor. The elasticity tensor is chosen accordingly to an isotropic homogeneous elastic material and the strains are assumed to be small. Furthermore, the pressure from the Reynolds equation is prescribed as a boundary condition for the stress field.

In order to establish a relation between the stresses in equation (5.1) and the dimensionless hydrodynamic pressure, the pressure boundary condition has to be transformed back to non-dimensionless form with

$$p = \underbrace{\frac{6\eta\Omega}{(c/R_i)^2}}_{B}\, p^{*}.$$

Since the radial clearance c appears not only in B but also in the transformation of the deformation to dimensionless form, the quantity $B^* = B\,c^2$ is varied to study the influence of viscosity and angular frequency. It is noted that B contains the angular frequency Ω instead of Ω_0, which is due to the fact that only steady state cases were considered for which $\Omega^* = 1$. The porous bush is chosen to have an elastic modulus of $E = 80\,\mathrm{GPa}$, a Poisson ratio $\nu = 0.3$, a dimensionless permeability $\Psi = 0.001$, an inner radius $R_i = 3\,\mathrm{mm}$ and an outer radius $R_o = 5.85\,\mathrm{mm}$. For $\eta = 0.01\ldots0.1\,\mathrm{Pa\,s}$ and $\Omega = 1000\ldots3000\,\mathrm{min}^{-1}$ results in the parameter range $B^* = 0.056\ldots1.696\,\mathrm{mN}$. The influence on the load capacity is shown in figure 5.12, where the relative deviation $\Delta = (F_{\mathrm{EHD}} - F_{\mathrm{HD}})/F_{\mathrm{HD}}$ versus B^* is shown. F_{EHD} refers to the load capacity which includes the bearing's deformation whereas F_{HD} represents the load capacity for a rigid bearing.

The greatest influence is found for a radial clearance of $c = 6\,\mu\mathrm{m}$ where for extreme cases ($B^* > 1.6\,\mathrm{mN}$ and $\varepsilon = 0.9$) a deviation from up to 45% is observed. This deviation occurs, however, for very high viscosity, eccentricity ratio and angular frequency. Under normal operating conditions the values of B^* will be significantly lower, probably around $B^* = 0.3\,\mathrm{MPa}$ for which the deviation lies around 17%. When the radial clearance is increased the influence diminishes rapidly with a maximum influence for $c = 13\,\mu\mathrm{m}$ and $\varepsilon = 0.9$ of about 10% even for the extreme cases. For all other investigated configurations the influ-

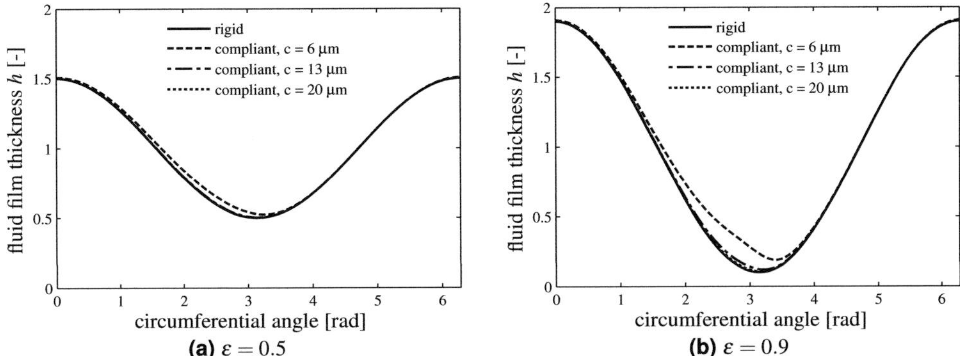

Figure 5.13: Elastic deformation of the dimensionless film thickness function in a porous journal bearing depending on the radial clearance and the eccentricity ratio for $B^* = 1.6\,\text{mN}$.

ence turns out to be significantly smaller if not negligible. It can therefore be concluded that the effect of elastic deformation does not have a significant influence on the load capacity under normal operating conditions for porous journal bearings with typical bearing geometries. For this reason the influence of elastic deformation of the porous bush in porous journal bearings is ignored in the proposed bearing model.

To illustrate the elastic deformation figure 5.13 shows the deformed dimensionless fluid film thickness function for different eccentricity ratios and radial clearance for extreme B^*. In accordance with the results from figure 5.12 the greatest deformation is observed for the compliant porous journal bearing with the smallest radial clearance. The largest deformation is situated where the fluid film thickness function converges. This also is where the pressure function normally reaches its maximum for steady states.

5.3 Solid journal bearing

By leaving out the radial term in equation (4.28) and setting the permeability to zero the Galerkin-based bearing model can also be applied to a solid journal bearing. The proposed model will be verified with the analytical short and long bearing solution and with a numerical discretization method, which were introduced in chapter 4. To compare the dynamical models

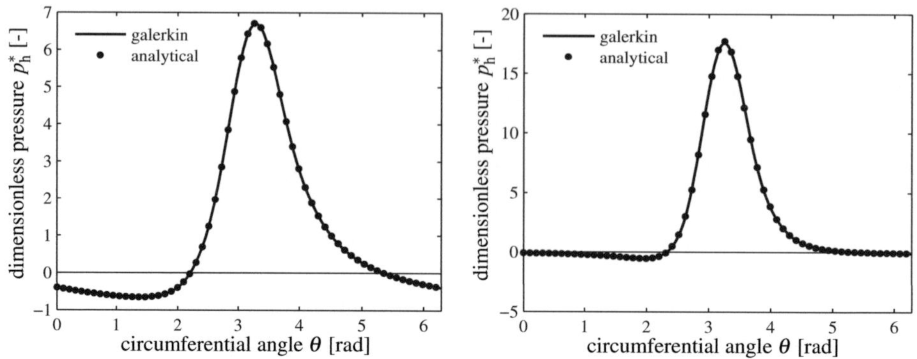

Figure 5.14: Comparisons of the pressure profile at $z^* = 0$ for the short (left) and long bearing (right) approximations of the Galerkin-based model and analytical solutions. Both pressure profiles correspond to the same dynamic case as in section 5.1 with $L/D = 1$ [151].

the same dynamical case is chosen as for the convergence study from section 5.1. All comparisons are made for a typical bearing geometry: a length to diameter ratio of $L/D = 1$ [151]. Comparisons with the analytical pressure functions for the short and long journal bearing approximations according to equation (4.22) and equation (4.23) respectively and according to the Galerkin-based model[1] are made in figure 5.14. Considering the small number of basis

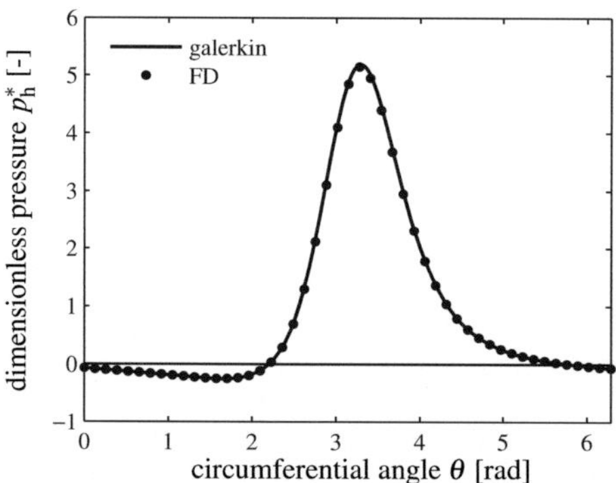

Figure 5.15: Comparison of the pressures profile at $z^* = 0$ of the Galerkin-based bearing model and a FD scheme for the finite bearing. This pressure profile corresponds to the same dynamic case as in section 5.1 with $L/D = 1$ [151].

---

[1]To allow a comparisons for the short and long bearing approximations, the same assumptions for the Reynolds equations were made for the Galerkin-based model (i.e. $\partial p_\mathrm{h}^*/\partial\theta \approx 0$ for the short bearing and $\partial p_\mathrm{h}^*/\partial z^* \approx 0$ for the long bearing approximation). Moreover, for the long bearing approximation the boundary condition $p_\mathrm{h}^*(\theta_1) = p_\mathrm{h}^*(\theta_1 + \pi) = 0$ is also prescribed for the Galerkin-based model. See section 4.2 for more details.

76

functions ($N = 10$, $M = 5$) both pressure curves and points are in excellent agreement.

Since no analytical solution exists for the finite journal bearing problem, we have to rely on a solution using another numerical discretization method to validate the Galerkin-based model for this case. Figure 5.15 shows a comparison between the numerical solution and the Galerkin-based model. The numerical solution was generated using the finite difference algorithm described in section 4.4 using the GÜMBEL cavitation condition. The same number of ansatz functions as for the previous examples was used for the model with the Galerkin-based model. Again, it can be concluded that both pressure profiles are in excellent agreement.

Figure 5.16 shows the dimensionless hydrodynamic load capacity versus the eccentricity ratio and the corresponding equilibrium points for the different bearing models. Figure 5.17 and figure 5.18 show the stiffness and damping coefficients respectively according to equation (4.36). All of these results were calculated at the equilibrium position corresponding to the eccentricity ratio (i.e. $\Omega^* = 1$). The short and long bearing approximations of the load capacity, equilibrium point and the stiffness and damping coefficients are in agreement with the analytical expressions from [152].

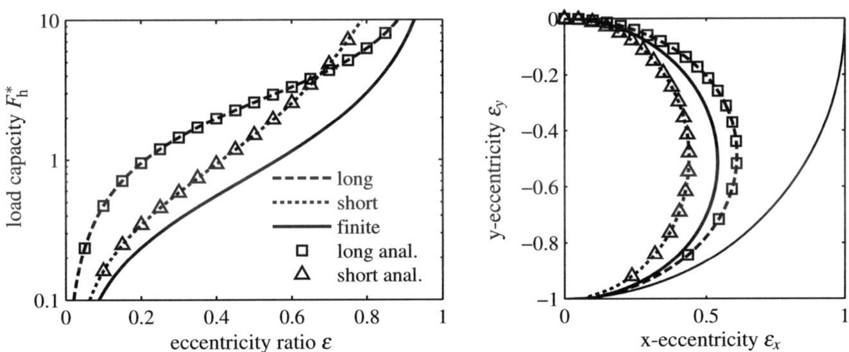

Figure 5.16: Dimensionless hydrodynamic load capacity (left) and equilibrium point (right) calculated with the Galerkin-based bearing models (short, long and finite) and comparisons with analytical solutions for the short and long bearing cases (short analytical and long analytical) [151].

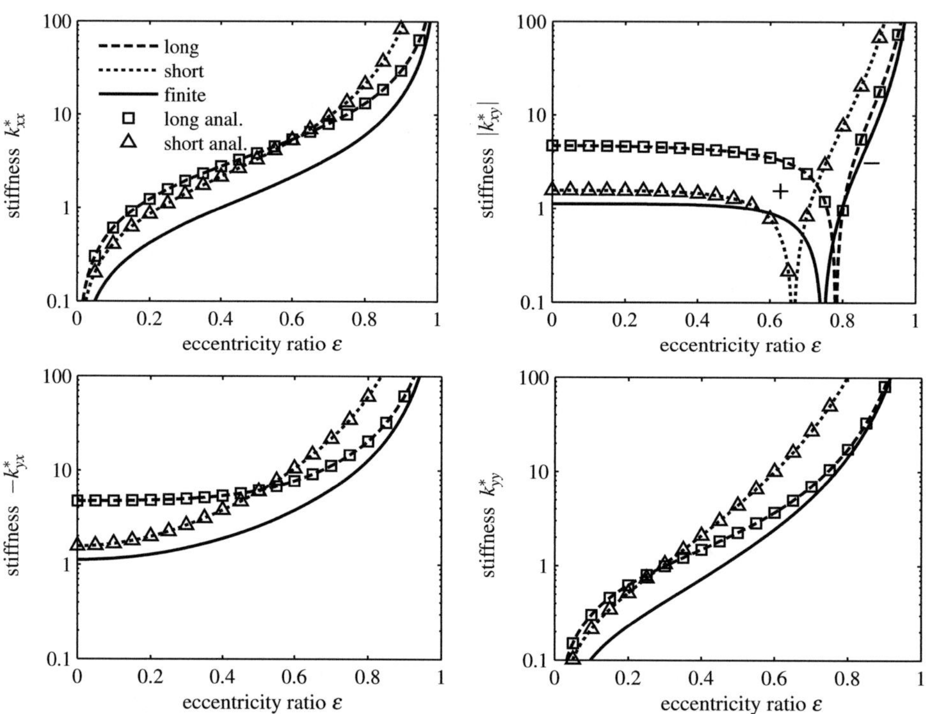

Figure 5.17: Dimensionless stiffness coefficients calculated with the Galerkin-based model (short, long and finite) and comparisons with analytical solutions for the short and long bearing cases (short analytical and long analytical). Note that the stiffness coefficient k_{xy}^* changes sign, as is indicated with $+$ and $-$ [151].

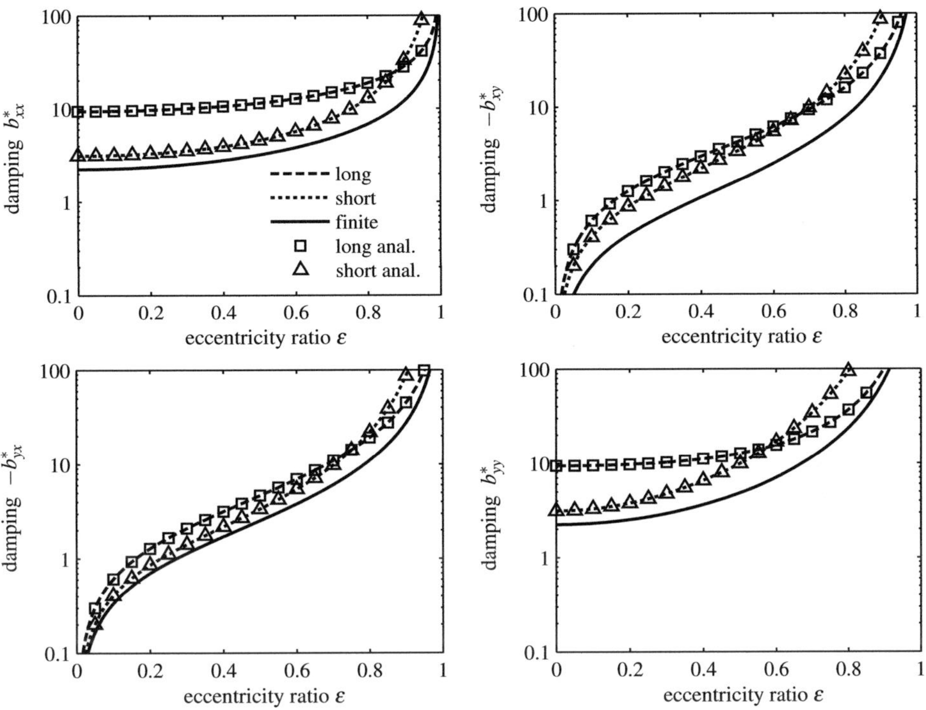

Figure 5.18: Damping coefficients calculated with the Galerkin-based model (short, long and finite) and comparisons with analytical solutions for the short and long bearing cases (short analytical and long analytical) [151].

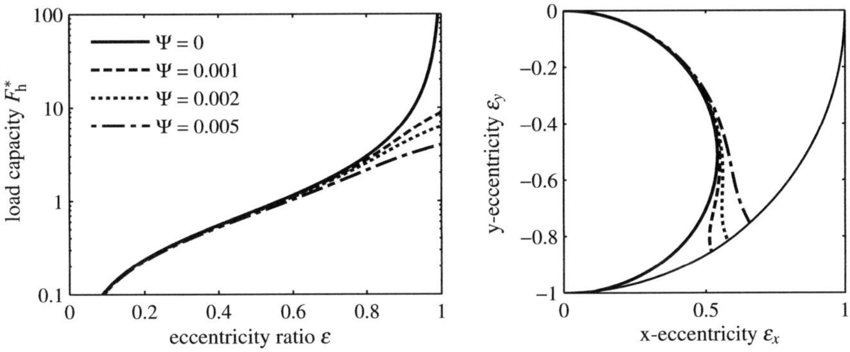

Figure 5.19: Hydrodynamic load capacity (left) and equilibrium point (right) for (porous) journal bearings. The case with $\Psi = 0$ represents the solid journal bearing [151].

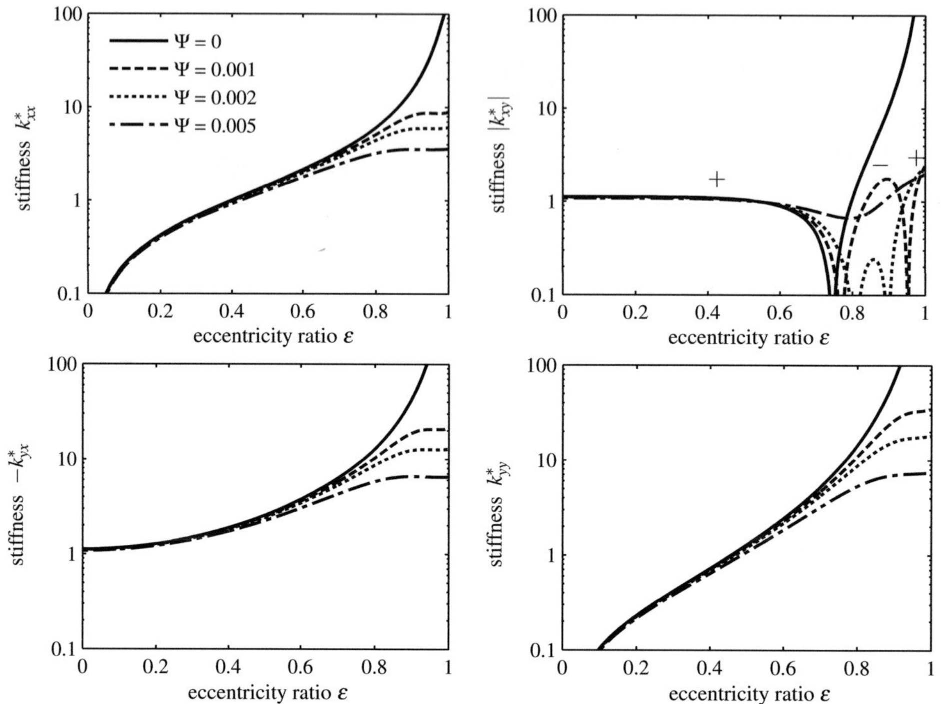

Figure 5.20: Dimensionless stiffness coefficients for (porous) journal bearings. The case with $\Psi = 0$ represents the solid journal bearing. Note that the stiffness coefficient k_{xy}^* for the classical and porous journal bearing changes sign, as is indicated with $+$ and $-$ [151].

5.4 Porous journal bearing

To solve the problem for a porous journal bearing Galerkin-based model with the unaltered ansatz function from equation (4.28) is now used. The same length to diameter ratio as for the previous examples will be used here, in addition the radial dimension of the porous bearing is defined by $R_o/R_i = 2$ [151].

Figure 5.19 shows the hydrodynamic load capacity and the equilibrium position of a solid journal bearing and three porous journal bearings. The stiffness and damping coefficients of three types of porous journal bearings in comparison with a solid journal bearing are shown in figure 5.20 and figure 5.21 respectively. It is clear that for higher eccentricity ratios (about $\varepsilon > 0.5$) the behavior of the porous journal bearing model differs significantly to that from the solid journal bearing. Not only the load capacity but also the damping and stiffness coefficients

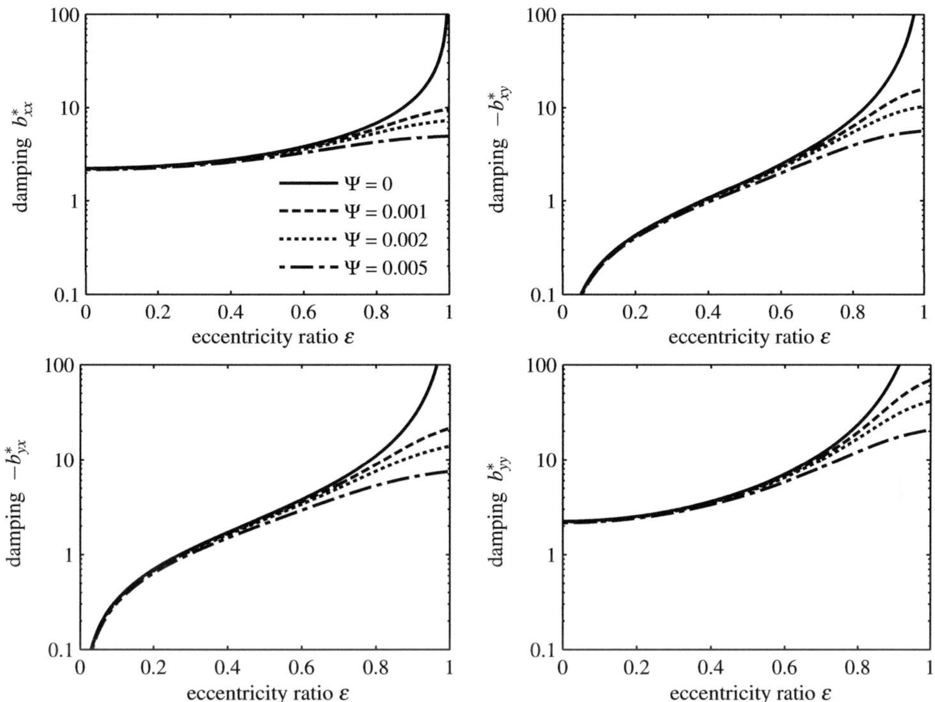

Figure 5.21: Dimensionless damping coefficients for (porous) journal bearings. The case with $\Psi = 0$ represents the solid journal bearing [151].

are lower compared to the solid journal bearing. Due to oil seepage into the porous material pressure build-up is significantly lower for higher eccentricity ratios, which influences the stiffness and damping properties of the fluid film. When the eccentricity ratio tends to one, the load capacity, the stiffness and the damping coefficients tend to infinity for the solid journal bearing, but approach a finite value for the porous bearings. This is also clearly seen in figure 5.22, where impedance maps for three different porous journal bearings are shown. In comparison to the impedance maps in figure 5.6 the magnitude of the impedance vector also approaches a finite value when the eccentricity ratio approaches one. From this it can be concluded that a given load will result in higher eccentricity ratios for porous journal bearings than for solid journal bearings.

Because of this it is much more likely that the surface roughness will start to noticeably influence the hydrodynamic pressure and that asperity contacts will generate additional contact pressure for porous journal bearings. The following will therefore firstly focus on the influ-

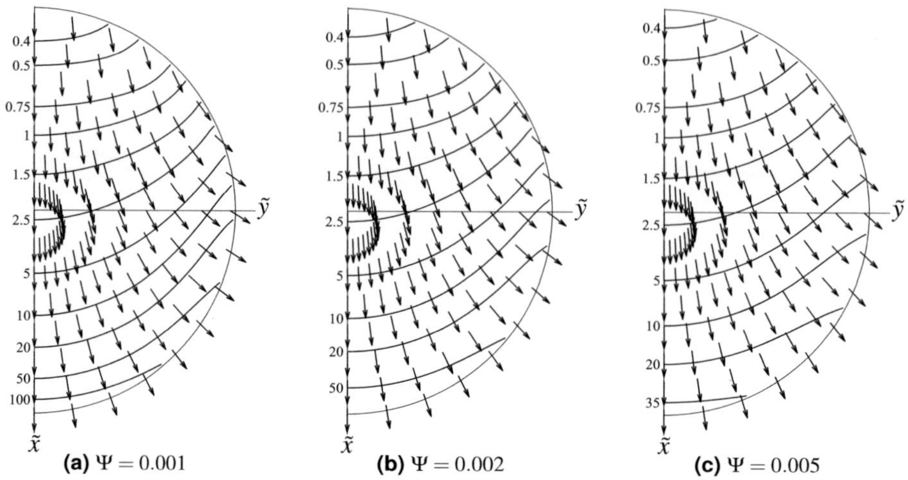

Figure 5.22: Impedance maps for porous journal bearings with $L/D = 1$, $R_o/R_i = 2$ and varying permeability Ψ.

ence of surface roughness on the hydrodynamic pressure and secondly investigate how asperity contacts affect the steady-state behavior of porous journal bearings.

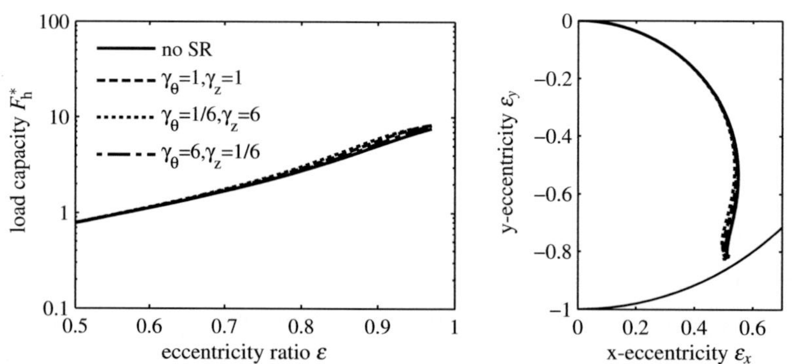

Figure 5.23: Hydrodynamic load capacity (left) and equilibrium point (right) for porous journal bearings (with $\Psi = 0.001$ and $R_o/R_i = 2$) with and without the influence of rough surfaces on the hydrodynamic pressure [151].

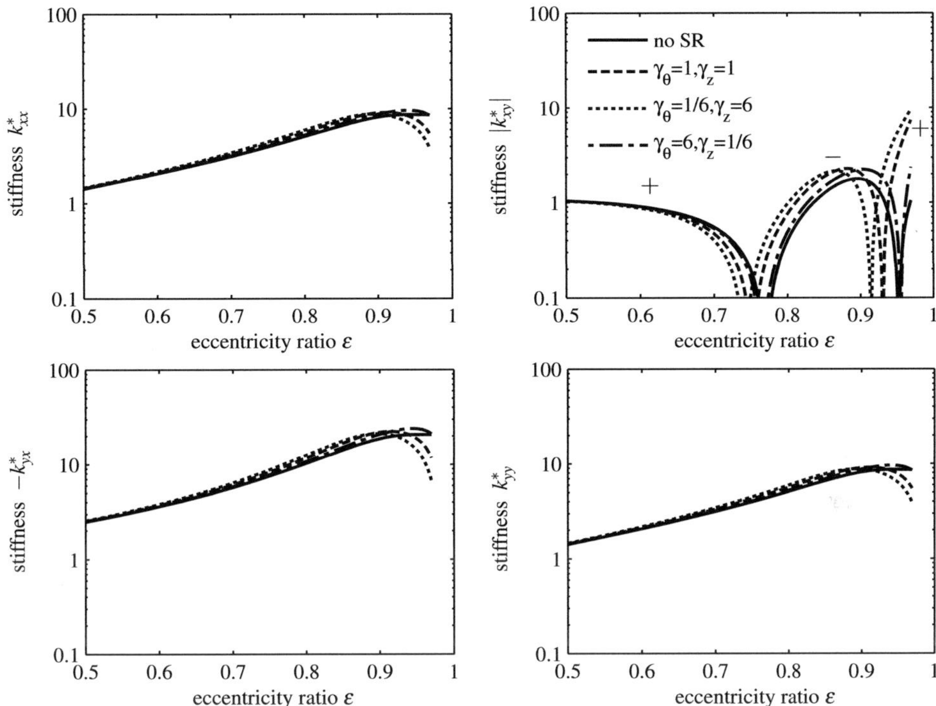

Figure 5.24: Dimensionless stiffness coefficients for porous journal bearings (with $\Psi = 0.001$ and $R_o/R_i = 2$) with and without the influence of rough surfaces on the hydrodynamic pressure. Note that the stiffness coefficient k_{xy}^* changes sign, as is indicated with $+$ and $-$ [151].

5.4.1 Rough surfaces

By modifying the weighted residual in equation (4.30) according to equation (4.43) the influence of rough surfaces on the hydrodynamical pressure is included into the model (although the influence of solid contact pressure is not yet considered). Due to additional diagonals in the system of linear equations that result from the added flow factors the solution procedure complicates slightly. However, the previously stated advantages of the Galerkin-based model still hold, the same numerical solver for the system of linear equations can be used and therefore the solution procedure remains efficient.

Results for three different roughness orientation scenarios (defined by the Peklenik number $\tilde{\gamma}$) from [124, 125] applied to a porous bearing are shown here: an isotropic roughness orientation ($\phi_\theta(\tilde{\gamma} = 1)$), a transversely orientated roughness ($\phi_\theta(\tilde{\gamma} = 1/6)$) and a longitudinally orientated roughness ($\phi_\theta(\tilde{\gamma} = 6)$). The reader is referred to figure 2.5 for a graphical representation of

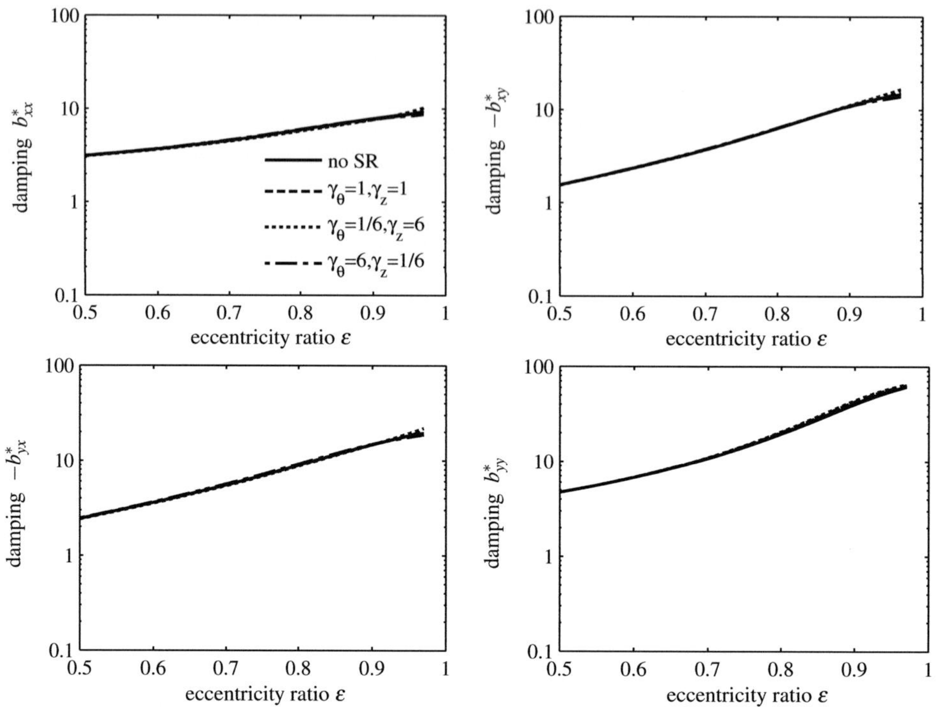

Figure 5.25: Dimensionless damping coefficients for porous journal bearings (with $\Psi = 0.001$ and $R_o/R_i = 2$) with and without the influence of rough surfaces [151].

these roughness profiles. The corresponding flow factors in axial direction are defined by the inverse Peklenik number: $\phi_z(\tilde{\gamma}) = \phi_\theta(1/\tilde{\gamma})$ and the Peklenik number for the shear flow factor is equal to the Peklenik number of the circumferential flow factor. The journal surface is assumed to be smooth and the ratio between the standard deviation of the bearing roughness and the radial clearance is chosen as $c/\sigma_h = 20$. The same bearing geometry as for the smooth porous journal bearings was used here and all porous bearing cases have a permeability of $\Psi = 0.001$ and an outer to inner bearing radius of $R_o/R_i = 2$ [151].

Figure 5.23 shows the hydrodynamic load capacity and the corresponding equilibrium points for a porous journal bearing with a smooth surface and three porous journal bearings with rough surfaces but different roughness orientations. Figure 5.24 and figure 5.25 show the corresponding stiffness and damping coefficients at equilibrium position respectively.

Due to the influence of surface roughness load capacity and stiffness coefficients turn out to be higher than for the smooth porous bearing, but only notably for higher eccentricity ratios

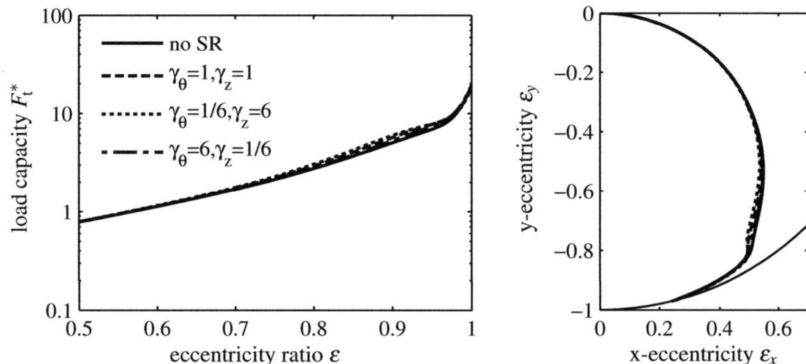

Figure 5.26: Total load capacity (left) and equilibrium point (right) for porous journal bearings (with $\Psi = 0.001$ and $R_o/R_i = 2$) with asperity contacts and with and without the influence of rough surfaces on the hydrodynamic pressure.

$(0.5 < \varepsilon < 0.9)$. Because of this lower eccentricity values are left out in figures 5.23-5.25. The transversely oriented roughness has the largest influence followed by the isotropic and the longitudinally oriented roughness. This trend observed for load capacity is also seen in [125]. The change for the damping coefficients seems negligible. This might be explained by the fact that the flow factors have no part in the time dependent term in the averaged Reynolds equation (see equation (4.43)).

For very high eccentricity ratios (about $\varepsilon > 0.9$) the influence of surface roughness seems to decrease the load capacity and stiffness coefficients. This effect is most prominently seen in the stiffness coefficients (excluding k_{xy}^*) for the transverse roughness case followed by the isotropic and longitudinally roughness case. Again this effect also is observed in [125] and can be explained with the change in slope of the shear flow factor. It should be mentioned that these slope changes might not be accurate due to large uncertainty in the original shear flow factors for small film thickness. Moreover do these results not include the influence of solid contact pressure which will play an increasingly important role in this film thickness regime [125].

It can be concluded that the influence of surface roughness on the hydrodynamical pressure has a small but notable influence on the bearing parameters studied here and that the proposed model is able to analyze this. Moreover, the Galerkin-based model is able to solve the problem without any significant additional numerical costs.

This work is mainly considered with the dynamical behavior of porous journal bearings within the context of a multibody system. In such an environment the deviations caused by the flow factors, which already were small, become even less significant. In addition the evaluation of flow factors requires complex numerical simulations to account for the micro-structure of the

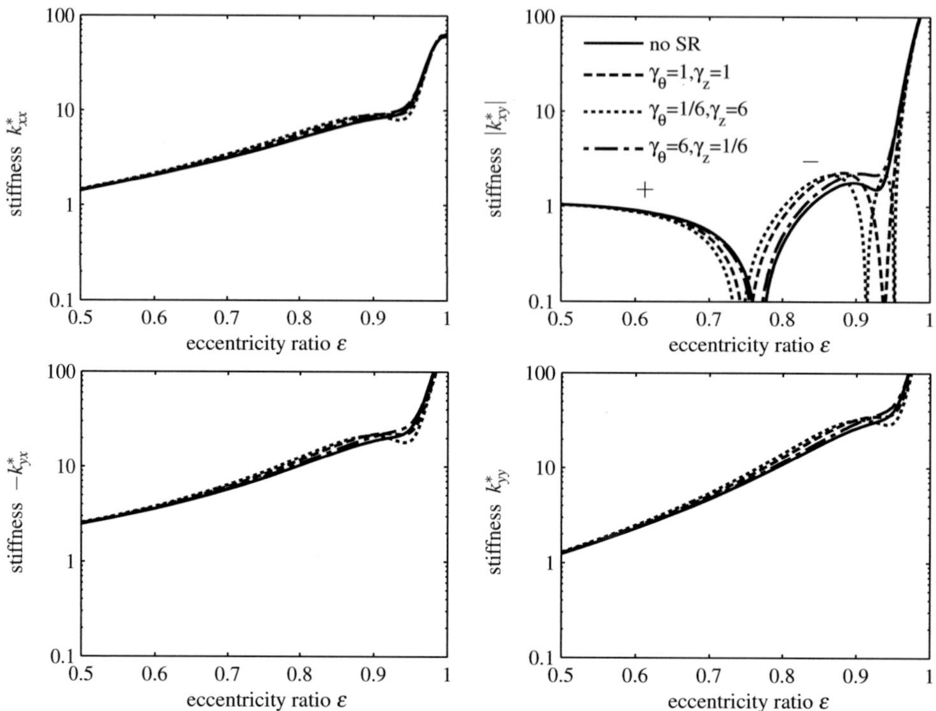

Figure 5.27: Dimensionless stiffness coefficients for porous journal bearings (with $\Psi = 0.001$ and $R_o/R_i = 2$) with asperity contacts and with and without the influence of rough surfaces on the hydrodynamic pressure. Note that the stiffness coefficient k^*_{xy} changes sign, as is indicated with $+$ and $-$.

surface, which is another practical issue. The influence of surface roughness on the hydrodynamical pressure is therefore not taken into account throughout the rest of this thesis. The next section features results including the influence of surface roughness through flow factors and asperity contacts to justify this.

5.4.2 Asperity contacts

Thus far all results only considered the hydrodynamic fluid pressure. Now the inclusion of contact pressure due to asperity contacts will be studied. The parameters to define the asperity contact are $c^{sr}_{GW} = 30$ and $c/\sigma_e = 50$ (see equation (4.12)). As for the previous results these parameters relate to a smooth to rough contact. The reader is reminded that the parameter σ_e relates to the distribution of the asperity peaks and not to the distribution of the original rough-

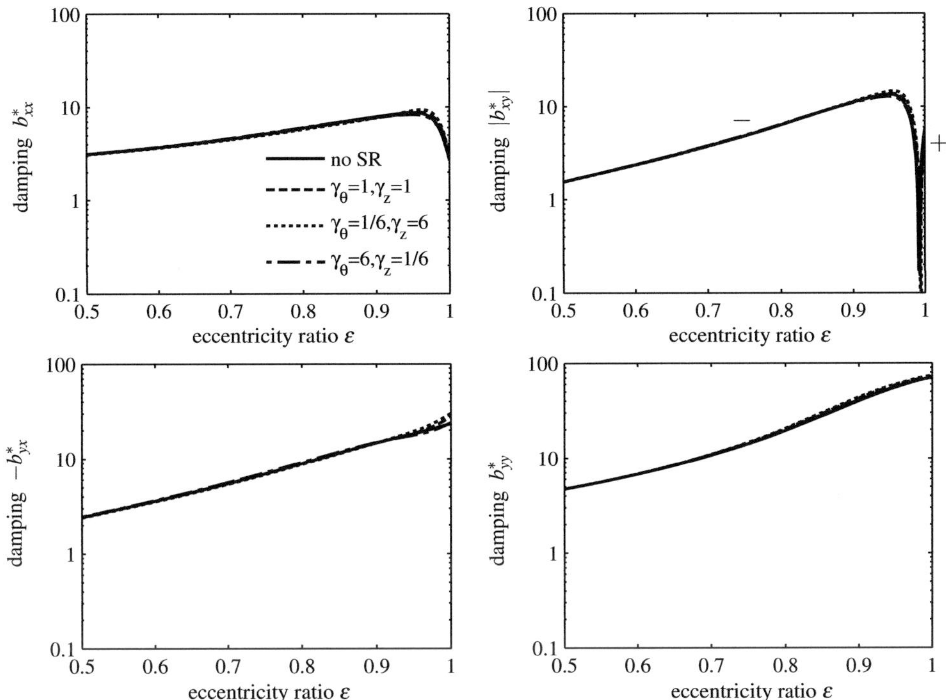

Figure 5.28: Dimensionless damping coefficients for porous journal bearings (with $\Psi = 0.001$ and $R_o/R_i = 2$) with asperity contacts and with/without the influence of rough surfaces on the hydrodynamic pressure. Note that the damping coefficient b_{xy}^* changes sign, as is indicated with $+$ and $-$.

ness profile. For the latter σ_h is of importance and for this the same value as in the previous section has been used. The standard deviation of the asperity peak heights σ_e is chosen such that it can be related to the same surface geometry which possesses the standard deviation σ_h. Figure 5.26 shows the total load capacity and equilibrium points for a porous journal bearing with the influence of the surface roughness on the hydrodynamic pressure and asperity contacts. The corresponding stiffness and damping coefficients are shown in figure 5.27 and figure 5.28. As expected an increase in the load capacity is observed for very high eccentricity ratios. The corresponding equilibrium position changes accordingly. The same effect is observed for the stiffness coefficients, but not for the damping coefficients. This is to be expected since the asperity contact force only depends on ε and not on ε'. Nevertheless, a sharp decrease is observed for b_{xx}^* and b_{xy}^*. This has to do with the fact the coefficients are observed in the $x-y$ coordinate system. The attitude angle is strongly influenced at high eccentricity ratios as is shown in figure 5.26 for the equilibrium point. This in turn also affects the horizontal bearing force $F_{t,x}^*$

as well as derived quantities. When the damping coefficients are observed in the coordinate system rotated by the attitude angle, these deviations do not occur.

For smaller eccentricity ratios the surface roughness only influences the hydrodynamical pressure, but for high eccentricities the asperity contact becomes influential. It is seen for all results that asperity contacts dominate for very small film thicknesses and the difference between the results with and without the influence of flow factors is very small. Only for the stiffness coefficient k_{xy}^* somewhat larger deviations are observed between $0.6 < \varepsilon < 0.9$, but these are still small compared to the influence of the asperity contact. It is also observed that k_{xy}^* for the isotropic and transversal roughness profiles briefly become positive again around $\varepsilon \approx 0.93$. For the other stiffness coefficients a small decrease is observed shortly before the asperity contacts become influential, which is most pronounced for the transversal roughness orientation $(\phi_\theta(\tilde{\gamma}=1/6))$. All of these deviations originate from the large uncertainty of the original shear flow factors for small film thickness from [125] and were also observed for the results without asperity contacts. Despite these differences, the influence of the current surface roughness on the hydrodynamical pressure is small and it is therefore justified to ignore its effects throughout the rest of this work.

5.5 Misalignment

To study the steady state characteristics of a misaligned plain journal bearing the influence of $\bar{\varepsilon}$ and ψ in equation (4.2) on the dimensionless hydrodynamical load capacity F_h^* and moment capacity M_h^* is investigated. The influence of asperity contacts has been excluded here. Contour plots of the load and moment capacity against ε and $\bar{\varepsilon}$ for $\psi = 0$ and $\psi = \pi/2$ are found in figures 5.29 and 5.30 for a solid journal bearing and in figures 5.31 and 5.32 for a porous journal bearing. $\psi = 0$ and $\psi = \pi/2$ represent the two extreme situation for the angle between the eccentricity vector and the journal axis projection on the mid-plane.

The load capacity of the solid journal bearing with $\bar{\varepsilon} = 0$ is in agreement with the results of the aligned solid journal bearing (see e.g. figure 5.19). When the journal axis projection $\bar{\varepsilon}$ is small the influence of misalignment on the load capacity seems dismissible especially for $\psi = \pi/2$. For larger $\bar{\varepsilon}$ an increase in load capacity is observed which is the largest for $\psi = 0$. Another important conclusion is that the load capacity seems to approach a finite value when the eccentricity ratio approaches its maximum value for non-zero $\bar{\varepsilon}$. As was already found in section 5.1 the load capacity actually approaches infinity, but only for very small minimum film thickness [15]. Even if the proposed bearing model would be able to predict this it would hardly be visible in the graph.

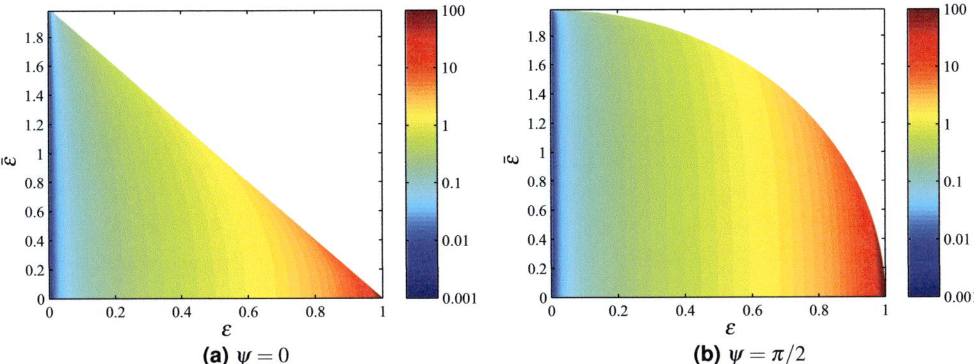

Figure 5.29: The dimensionless hydrodynamical load capacity F_h^* against ε and $\bar{\varepsilon}$ for different ψ for a misaligned solid journal bearing with $L/D = 1$.

It is clear that the moment capacity has to vanish when $\bar{\varepsilon} = 0$. The moment capacity increases when the eccentricity ratio ε and/or the journal axis projection on the mid-plane $\bar{\varepsilon}$ are increased. The reader is reminded that theoretically also here the moment capacity approaches infinity when the minimum film thickness goes to zero.

When looking at the results for porous journal bearings a similar behavior is observed as for the solid journal bearings, but with smaller load and moment capacity when the minimum film thickness is approached. An interesting difference is that for porous journal bearings the maximum moment capacity is now reached for much smaller eccentricity ratios as for conventional journal bearings.

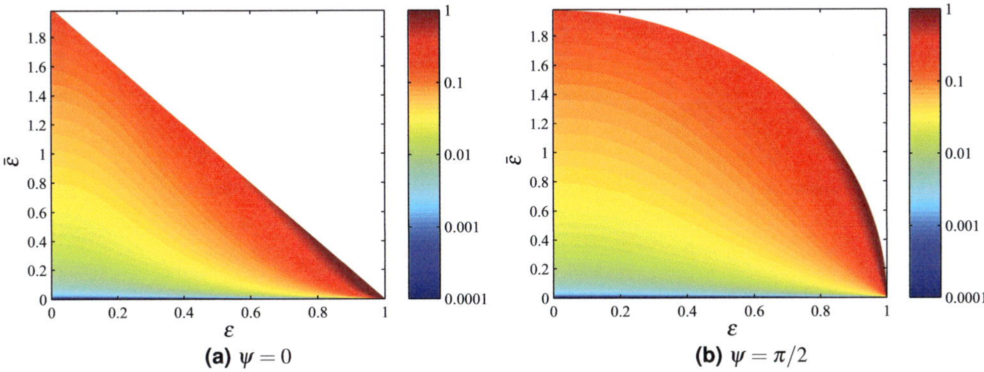

Figure 5.30: The dimensionless hydrodynamical moment capacity M_h^* against ε and $\bar{\varepsilon}$ for different ψ for a misaligned solid journal bearing with $L/D = 1$.

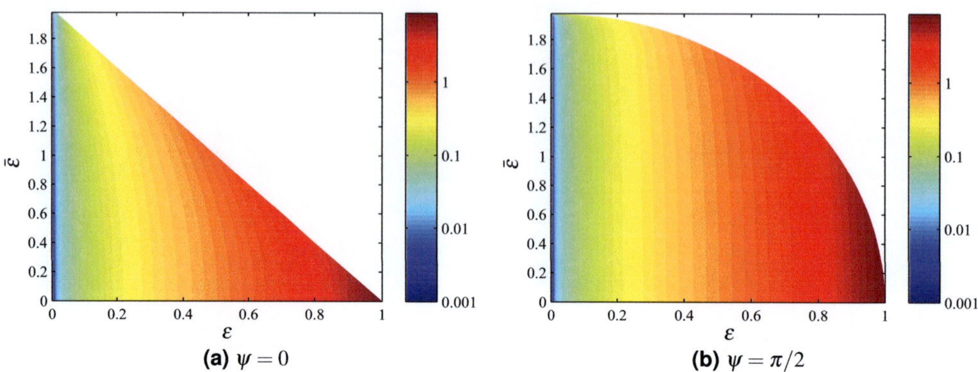

(a) $\psi = 0$ **(b)** $\psi = \pi/2$

Figure 5.31: The dimensionless hydrodynamical load capacity F_h^* against ε and $\bar{\varepsilon}$ for different ψ of a misaligned porous journal bearing with $L/D = 1$, $\Psi = 0.001$ and $R_o/R_i = 2$.

In general it can be concluded that the implications of misalignment on the load capacity are less significant for large and moderate film heights. When the film thickness function becomes relatively small the load capacity can be significantly worse than for aligned bearings. However, under normal operating conditions and moderate film thickness misalignment will most likely become influential through the additional moment capacity.

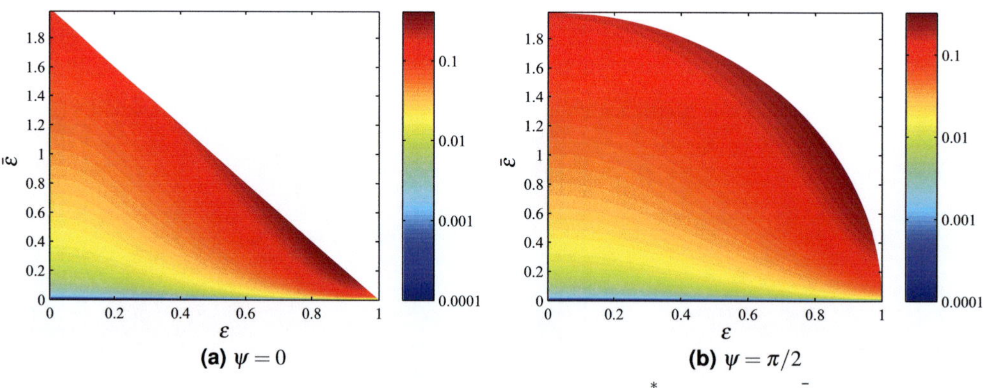

(a) $\psi = 0$ **(b)** $\psi = \pi/2$

Figure 5.32: The dimensionless hydrodynamical moment capacity M_h^* against ε and $\bar{\varepsilon}$ for different ψ of a misaligned porous journal bearing with $L/D = 1$, $\Psi = 0.001$ and $R_o/R_i = 2$.

Chapter 6

Rotor-bearing systems containing plain journal bearings

The proposed bearing model will now be applied in the context of several exemplary rotor-bearing systems containing plain journal bearings. For such systems, the stability of their stationary solution is investigated. The existence of fixed-point and periodic solutions are examined by following the solutions for changing system parameters along with the detection of any bifurcation points. To investigate the dynamical behavior of such systems the freely available toolbox MATCONT for MATLAB has been used [39]. The studies for rotor-bearing systems were performed on rotors with and without unbalance, with porous and solid journal bearings both including for asperity contacts. The first section features systems with rigid rotors, followed by results of a rotor-bearing system with a flexible rotor where misalignment for the bearings is taken into account. Finally, a short description of the inclusion of the proposed Galerkin-based bearing model into multibody system software is given and an example of a multibody system is presented to which the approach was applied.

6.1 Elementary rotor-bearing systems with rigid rotors

In this section elementary rotor-bearing systems will be studied consisting of one rotor mounted by two bearings. The rotor has mass m, an optional unbalance a and is loaded by a force F_0 as is illustrated in figure 6.1. The parameter to scale the angular frequency is defined as $\Omega_0 = \sqrt{F_o/mc}$ analogue to [152]. By assuming symmetry both bearings behave identical and only one set of eccentricity coordinates has to be considered[1]. All bearings have a length to diameter ratio of $L/D = 1$, the bearing's radial thickness is defined with $R_o/R_i = 2$ and its permeability is varied. The journal's surface is assumed smooth whereas the bearing's surface roughness is retained. With this assumption the parameters for the asperity contacts are chosen as $c_{GW}^{sr} = 30$ and $c/\sigma_e = 50$ similar to the asperity contacts used in section 5.4.2.

For a rigid rotor with two identical plain journal bearings perfect alignment of the bearing's axis with respect to the rotor's axis is assumed. The equations of motion in dimensionless form (see section 3.1.2) for this rotor-bearing system read

$$\begin{cases} \varepsilon_x'' = 2F_{t,x}^*/F_0^* + \Omega^{*2}a^*\cos(\Omega^*\tau) \\ \varepsilon_y'' = 2F_{t,y}^*/F_0^* - 1 + \Omega^{*2}a^*\sin(\Omega^*\tau) \end{cases} , \tag{6.1}$$

where $a^* = a/c$ is the dimensionless unbalance and $F_{t;x,y}^*$ are the bearing forces calculated with equation (4.10). Due to the introduced unbalance the system depends explicitly on time, which puts it into the category of non-autonomous dynamical systems. By expressing the harmonic excitation as a limit cycle of a separate autonomous oscillatory system, the equations of motion can be stated in autonomous form:

$$\begin{cases} u_1' = u_2 \\ u_2' = 2F_{t,x}^*/F_0^* + \Omega^{*2}a^*u_6 \\ u_3' = u_4 \\ u_4' = 2F_{t,y}^*/F_0^* - 1 + \Omega^{*2}a^*u_5 \\ u_5' = u_5 + \Omega^*u_6 - u_5(u_5^2 + u_6^2) \\ u_6' = u_6 - \Omega^*u_5 - u_6(u_5^2 + u_6^2) \end{cases} . \tag{6.2}$$

With equation (6.2) the phase-space dimensions are given by $u_1 = \varepsilon_x$, $u_2 = \varepsilon_x'$, $u_3 = \varepsilon_y$, $u_4 = \varepsilon_y'$, $u_5 = \sin(\Omega^*\tau)$ and $u_6 = \cos(\Omega^*\tau)$. The last two phase-space dimensions correspond to the added nonlinear harmonic oscillator to avoid non-autonomy [40]. The system is formulated in autonomous form to allow an implementation in MATCONT, which only supports autonomous systems.

---

[1]It should be noted that by making this assumption certain rigid body modes such as conical modes cannot be studied.

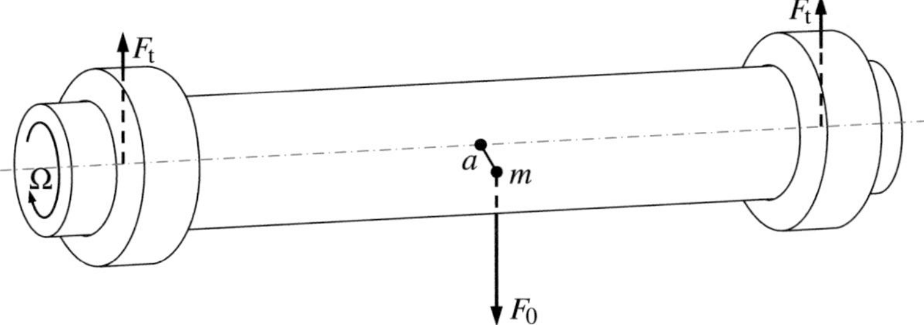

Figure 6.1: The symmetric rotor-bearing system with a rigid rotor (mass m), an unbalance a and an applied force F_0.

To optimize robustness and performance for stiff problems MATCONT offers the possibility to prescribe the Jacobian matrix in explicit form. For the current phase-space this is given by

$$
J(u) = \begin{pmatrix}
0 & 1 & 0 & 0 & 0 & 0 \\
-2k_{xx}^*/F_0^* & -2b_{xx}^*/F_0^* & -2k_{xy}^*/F_0^* & -2b_{xy}^*/F_0^* & 0 & \Omega^{*2}a^* \\
0 & 0 & 0 & 1 & 0 & 0 \\
-2k_{yx}^*/F_0^* & -2b_{yx}^*/F_0^* & -2k_{yy}^*/F_0^* & -2b_{yy}^*/F_0^* & \Omega^{*2}a^* & 0 \\
0 & 0 & 0 & 0 & 1-3u_5^2-u_6^2 & \Omega^*-2u_5u_6 \\
0 & 0 & 0 & 0 & -\Omega^*-2u_5u_6 & 1-u_5^2-3u_6^2
\end{pmatrix},
\tag{6.3}
$$

where the stiffness and damping coefficients k_{ij}^* and b_{ij}^* are calculated using the methods introduced in section 4.3.1. It is noted that, contrary to the steady state coefficients shown in the previous chapter (which only depend on ε_x and ε_y), these coefficients also depend on the translational velocities ε_x' and ε_y'. Similar to the usual Jacobian in addition a Jacobian matrix with respect to the changing parameter(s) is needed, which is in this case solely the angular frequency Ω^*:

$$
J_p(u) = \begin{pmatrix}
0 \\
2/F_0^* \, \partial F_x^*/\partial\Omega^* + 2\Omega^*a^*u_6 \\
0 \\
2/F_0^* \, \partial F_y^*/\partial\Omega^* + 2\Omega^*a^*u_5 \\
u_6 \\
-u_5
\end{pmatrix}.
\tag{6.4}
$$

6.1.1 Equilibrium stability

For balanced rotor-bearing systems (i.e. the system from equation (6.2) with $a^* = 0$) the local stability of its fixed-point solutions can be determined using linear stability analysis. The stability limits of fixed-point solutions can be obtained by observing the eigenvalues of the Jacobian matrix (see section 3.2).

This can be done by numerically calculating the Jacobian matrix. Alternatively, the steady-state stiffness and damping coefficients evaluated in the previous chapter can now be utilized to construct the Jacobian matrix. Subsequently, its characteristic equation can be analyzed using the Routh-Hurwitz stability criterion, which provides conditions that have to be satisfied for the solution to be stable [140]. For $a^* = 0$ the characteristic equation is of the fourth order: $a_4 s^4 + a_3 s^3 + a_2 s^2 + a_1 s + a_0 = 0$ for which the Routh-Hurwitz criterion prescribes the following stability conditions: $a_i > 0$, $a_3 a_2 > a_4 a_1$ and $a_3 a_2 a_1 > a_4 a_1^2 + a_3^2 a_0$. The first two conditions are always satisfied, the last condition eventually yields

$$\Omega^* F_0^* < \frac{2\,\tilde{a}_1 \tilde{a}_{22} \tilde{a}_3}{\tilde{a}_0 \tilde{a}_3^2 + \tilde{a}_1^2 - \tilde{a}_1 \tilde{a}_{21} \tilde{a}_3}, \tag{6.5}$$

where

$$\begin{aligned}
\tilde{a}_0 &= k_{xx}^* k_{yy}^* - k_{xy}^* k_{yx}^* \\
\tilde{a}_1 &= b_{xx}^* k_{yy}^* + b_{yy}^* k_{xx}^* - b_{xy}^* k_{yx}^* - b_{yx}^* k_{xy}^* \\
\tilde{a}_{21} &= k_{xx}^* + k_{yy}^* \\
\tilde{a}_{22} &= b_{xx}^* b_{yy}^* - b_{xy}^* b_{yx}^* \\
\tilde{a}_3 &= b_{xx}^* + b_{yy}^*.
\end{aligned}$$

Figure 6.2 shows the critical rotor frequency Ω_{is}^* versus the eccentricity ratio ε and versus the total applied load F_0^* for a solid journal bearing ($\Psi = 0$) and several porous journal bearings ($\Psi \neq 0$) in the hydrodynamical regime.

The critical rotor frequency of the rotor-bearing system in general decreases when the permeability increases. Theoretically, the solid journal bearing will always be stable when $\varepsilon > 0.82$, whereas for the porous journal bearings this happens when the asperity contacts become influential for $\varepsilon > 0.94$. The critical rotor frequency in the mixed lubrication regime is shown in figure 6.3 and illustrates these limits.

Generally, the critical rotor frequency increases when the eccentricity ratio and the applied load are increased. However, for the porous journal bearings a decrease is observed shortly before

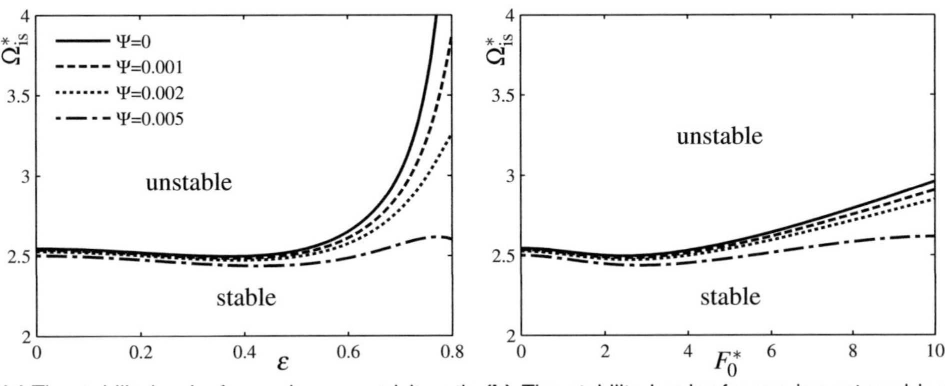

(a) The stability border for varying eccentricity ratio ε.

(b) The stability border for varying external load F_0^*.

Figure 6.2: The stability border for equilibrium points of a rigid rotor symmetrically mounted by aligned porous journal bearings in the hydrodynamic lubrication regime for various values of the dimensionless permeability factor Ψ.

the asperity contacts become influential. This slope change can directly be related to the slope change of the stiffness coefficient k_{xy}^* (see figure 5.20).

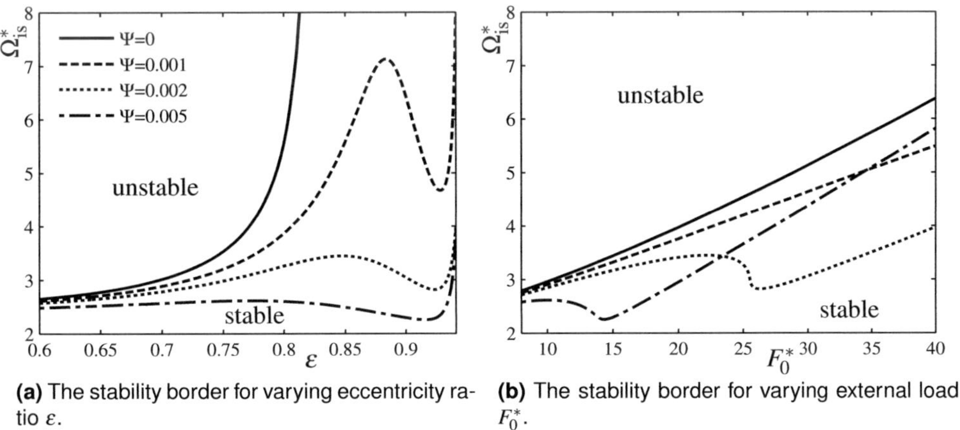

(a) The stability border for varying eccentricity ratio ε.

(b) The stability border for varying external load F_0^*.

Figure 6.3: The stability border for equilibrium points of a rigid rotor symmetrically mounted by aligned porous journal bearings in the mixed lubrication regime for various values of the dimensionless permeability factor Ψ.

6.1.2 Bifurcation analysis

The bifurcation behavior of the symmetric rotor-bearing system described in equation (6.2) with varying load F_0^*, permeability Ψ and unbalance a^* is studied using MATCONT [39]. As is the case for all bifurcation diagrams, stable solution branches are represented by solid lines and unstable solution branches by dashed lines. The scalar measure for the bifurcation diagram to describe the solutions is the maximum eccentricity ratio of the fixed-point or periodic solution for a given operating point. Bifurcation points are marked with different icons explained in the legend of each figure.

Solid journal bearing

Figure 6.4 shows bifurcation diagrams for a solid journal bearing with varying unbalance and $F_0^* = 1$. The equilibrium solution of the balanced rotor-bearing system remains stable until a sub-critical Hopf bifurcation is encountered at $\Omega^* = 2.526$. The same result is obtained by means of linear stability analysis which predicts a stable system below this angular frequency and for $\varepsilon < 0.17$. From this Hopf bifurcation an additional unstable branch with periodic solutions emanates and the fixed-point solution continues as an unstable fixed-point solution. The branch with periodic solutions emanating from the Hopf point becomes stable again at a fold bifurcation at $\Omega^* = 2.443$. It remains stable until at $\Omega^* = 7.116$ the continuation algorithm fails to follow the periodic solution due to numerical issues. As is indicated with the circled numbers, this branch contains period-1 solutions. The origin of the Hopf bifurcation can be attributed to self-excited vibrations originating from the journal bearings. The frequency of the periodic solution approximately equals half the angular frequency of the rotor and therefore is known $\frac{1}{2}\Omega$-whirl [152]. Moreover is it interesting to note that the sub-critical nature of this bifurcation results in a jump for the journal's position when the bifurcation is passed.

For $a^* = 0.1$ the system already has a period-1 solution due to the unbalance. This solution is stable until a sub-critical period doubling bifurcation occurs at $\Omega^* = 2.208$ after which it becomes unstable. Its cause is the same as the previously found Hopf bifurcation, but now the unbalance and the journal bearing's self-excited vibrations result in a branch with period-2 solutions. The solutions are initially unstable and eventually become stable at a fold bifurcation at $\Omega^* = 2.184$. This solution branch remains stable at least until $\Omega^* = 20$, after which the continuation was aborted. Another period doubling occurs at $\Omega^* = 3.055$ from which a branch of unstable period-2 solutions emanates. The solutions of this branch remain unstable at least until $\Omega^* = 3.242$, where the continuation failed due to numerical issues. Along the

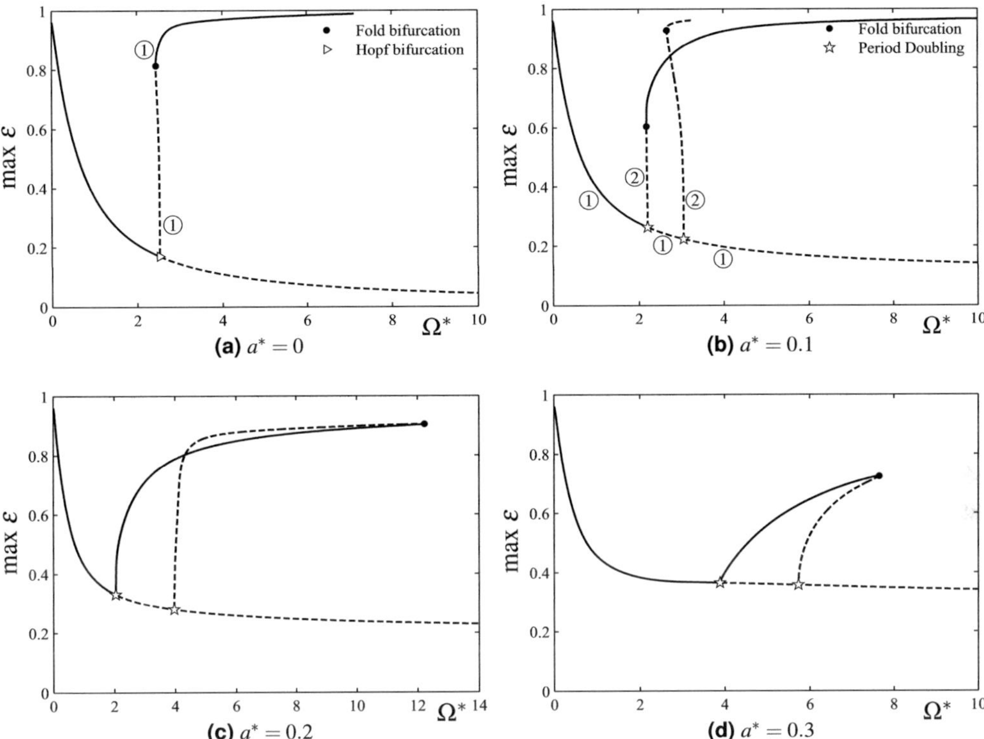

Figure 6.4: Bifurcation diagrams for a symmetric rigid rotor-bearing system containing solid journal bearings with $L/D = 1$, $F_0^* = 1$ and with varying unbalance a^*. The circled numbers indicate the periodicity of a branch with periodic solutions.

branch a fold bifurcation took place due to the fact that an additional Floquet multiplier left the unit circle on the complex plane. For clarity, the periodicity of the solutions belonging to the branches around the period doubling bifurcation is indicated with circled numbers. All investigated rotor-bearing systems with unbalance and solid journal bearings show the same periodicity around these period doubling bifurcations.

Increasing the unbalance to $a^* = 0.2$ has the effect that the period doubling bifurcation observed for the previous case now already takes place at $\Omega^* = 2.045$ and is super-critical. The initially stable period-2 solutions from the emanating branch become unstable at a fold bifurcation at $\Omega^* = 12.23$. Another branch of unstable period-2 solutions emanates from a period doubling bifurcation at $\Omega^* = 3.97$ along the branch with period-1 solutions. Eventually, this branch coalesces with the other period-2 branch at the fold bifurcation at $\Omega^* = 12.23$. It should

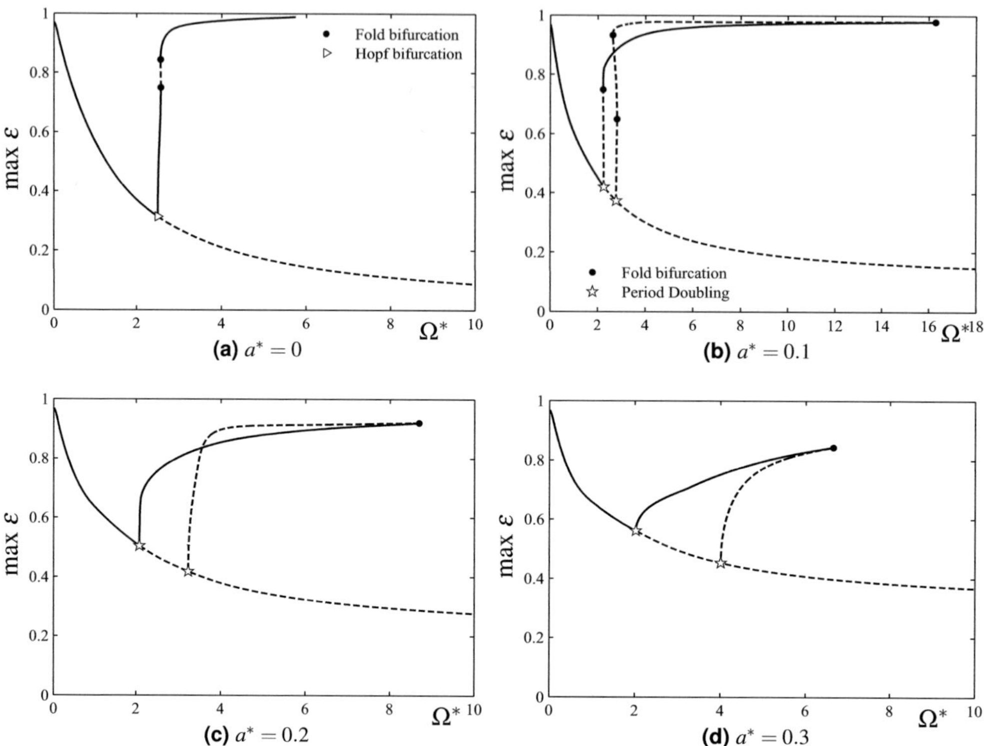

Figure 6.5: Bifurcation diagrams for a symmetric rigid rotor-bearing system containing solid journal bearings with $L/D = 1$, $F_0^* = 2$ and with varying unbalance a^*.

be noted that for larger rotor speeds not only unstable period-1 solutions exist, but that also quasi-periodic or chaotic solutions can coexist here. The numerical continuation algorithm is not able to detect these solutions, however, these can be visualized for chosen parameters by means of numerical integration. As was found, such solutions occur for almost all treated symmetric rigid rotor-bearing systems with unbalance including those containing porous journal bearings, which will be introduced later on. In retrospect, for $a^* = 0.1$, the same observed bifurcation scenario probably occurs as for $a^* = 0.2$, although it could not be detected due to numerical issues.

Another qualitatively similar case is observed when $a^* = 0.3$. However, now the first supercritical period doubling bifurcation occurs at $\Omega^* = 3.887$. It is also noted that the maximum eccentricity ratios for the periodic solutions turn out to be much smaller than for the previous case. The emanating stable solution branch now becomes unstable at a fold bifurcation at

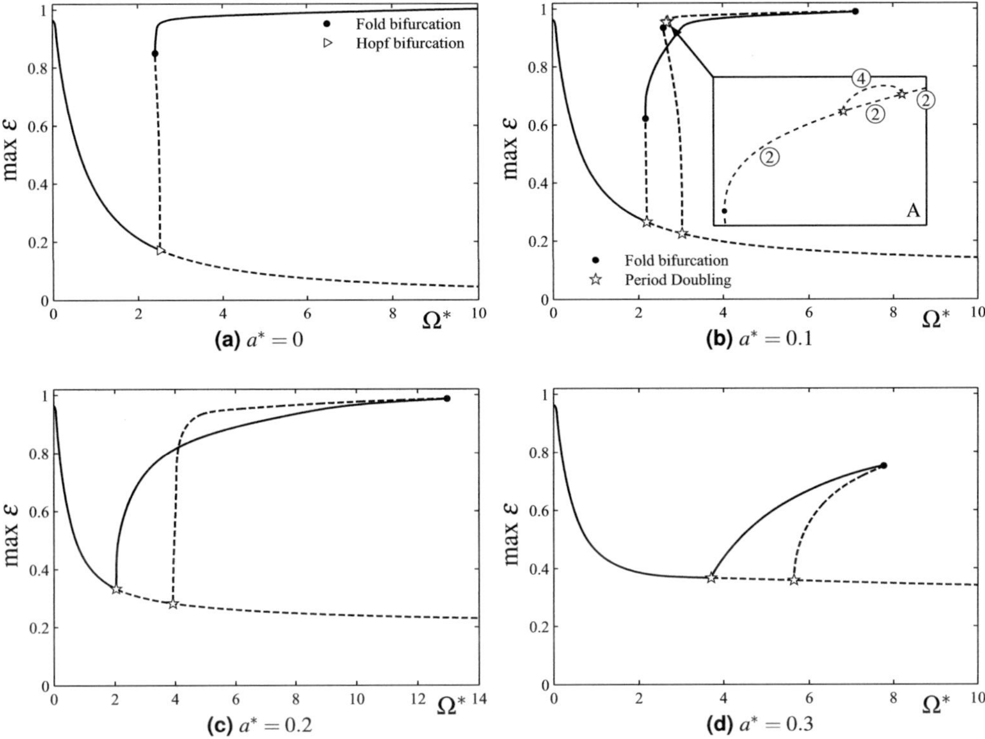

Figure 6.6: Bifurcation diagrams for a symmetric rigid rotor-bearing system containing porous journal bearings with $L/D = 1$, $\Psi = 0.001$, $F_0^* = 1$ and with varying unbalance a^*. The circled numbers indicate the periodicity of a branch with periodic solutions.

$\Omega^* = 7.659$, where it coalesces with a branch of unstable period-2 solutions emanating from the period doubling bifurcation at $\Omega^* = 5.734$. When observing the solutions beyond the fold bifurcation, quasi-periodic solutions are encountered. The reader is referred to figure 3.3c, which illustrates the solution at $\Omega^* = 10$.

The influence of load for a solid journal bearing is investigated in figure 6.5, where the applied load is raised to $F_0^* = 2$. For a balanced rotor-bearing system the stable fixed-point solution becomes unstable after a Hopf bifurcation is observed for $\Omega^* = 2.5$. The same onset of instability is found in figure 6.2 for $F_0^* = 2$ when $\varepsilon < 0.31$. The Hopf bifurcation is now super-critical instead of sub-critical (as was the case for $F_0^* = 1$). Thus, an emanating branch of stable periodic solutions is observed, whereas the branch of fixed-point solutions loses its stability after the Hopf bifurcation. The periodic solutions in this branch briefly become unstable between two

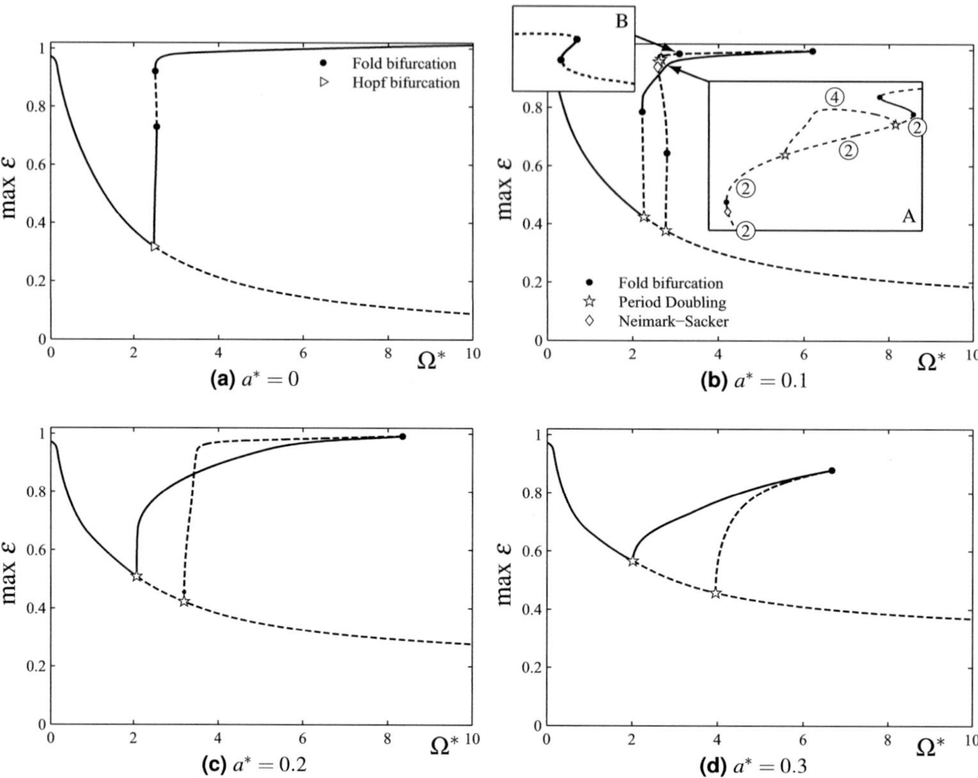

Figure 6.7: Bifurcation diagrams for a symmetric rigid rotor-bearing system containing porous journal bearings with $L/D = 1$, $\Psi = 0.001$, $F_0^* = 2$ and with varying unbalance a^*. The circled numbers indicate the periodicity of a branch with periodic solutions.

fold bifurcations at $\Omega^* = 2.57$ and $\Omega^* = 2.564$ where one of the multipliers has a magnitude larger than one. After this the solution remains stable until $\Omega^* = 5.741$ where the continuation was aborted by MATCONT due to numerical convergence issues.

For $a^* = 0.1$ the initial stable periodic solution becomes unstable at a sub-critical period doubling bifurcation at $\Omega^* = 2.27$. Here a branch of unstable period-2 solutions originates, which eventually becomes stable at a fold bifurcation at $\Omega^* = 2.237$ and remains stable until another fold bifurcation at $\Omega^* = 16.28$ is encountered. The path-follower fails to follow the emanating branch of unstable solutions at $\Omega^* = 3.328$. Another period doubling occurs at $\Omega^* = 2.789$, from which also a branch of unstable solutions emanates. Along this branch the solution gains and loses a stable Floquet multiplier at fold bifurcations at $\Omega^* = 2.821$ and $\Omega^* = 2.641$ respectively. Finally, also here MATCONT fails to follow the solution at $\Omega^* = 3.271$. It is believed to

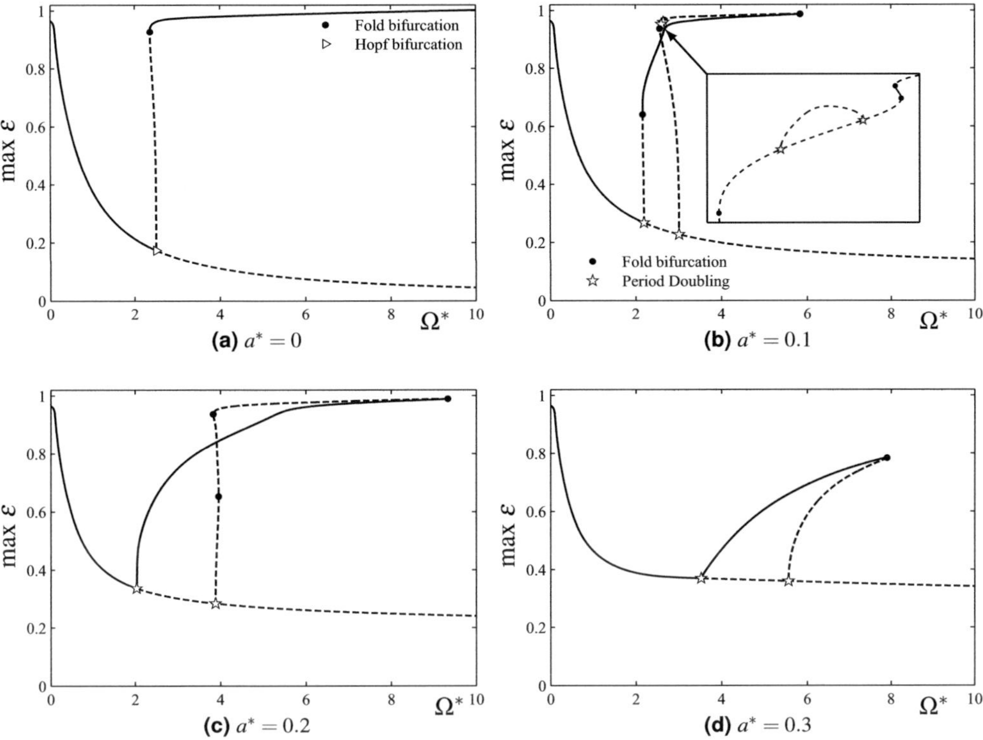

Figure 6.8: Bifurcation diagrams for a symmetric rigid rotor-bearing system containing porous journal bearings with $L/D = 1$, $\Psi = 0.002$, $F_0^* = 1$ and with varying unbalance a^*.

join up with the other branch of period-2 solutions, but this could not be verified.

Increasing a^* produces similar results as for the corresponding cases with $F_0^* = 1$. Two period doubling bifurcation are observed at $\Omega^* = 2.079$ and $\Omega^* = 3.214$ for $a^* = 0.2$ and at $\Omega^* = 2.024$ and $\Omega^* = 4.01$ for $a^* = 0.3$. From the first period doubling bifurcation a stable branch originates to become unstable at a fold bifurcation at $\Omega^* = 8.69$ for $a^* = 0.2$ and $\Omega^* = 6.649$ for $a^* = 0.3$ before an emanating branch of unstable solutions returns to the second period doubling bifurcation.

It can be concluded that higher unbalance generally decreases the onset frequency of instability (i.e. at which frequency the first Hopf or period doubling bifurcation is observed). However, depending on the applied load this does not always hold as if for example the case for $a^* = 0.3$. Raising the applied load will of course result in higher maximum eccentricity ratios. Lower loads on the other hand will accomplish that the emanating branch with period-1 ($a^* = 0$) or

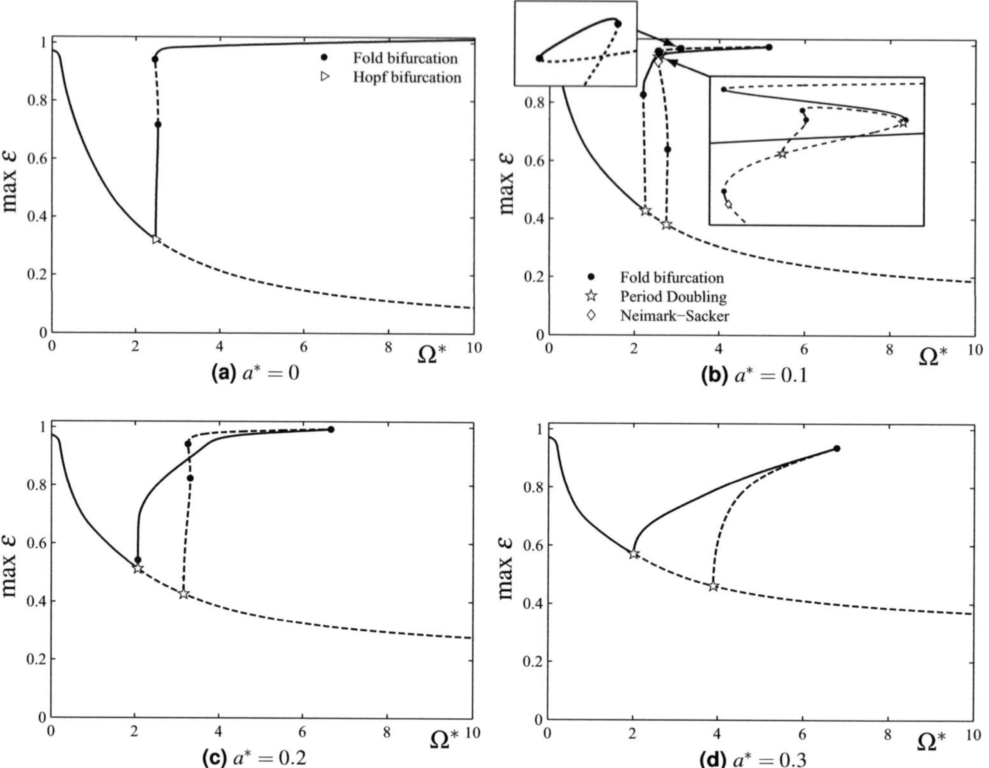

Figure 6.9: Bifurcation diagrams for a symmetric rigid rotor-bearing system containing porous journal bearings with $L/D = 1$, $\Psi = 0.002$, $F_0^* = 2$ and with varying unbalance a^*.

period-2 ($a^* \neq 0$) solutions maintains its stability up to a higher rotor speed.

The witnessed numerical convergence issues are directly related to the accuracy demands that become problematic for very high eccentricity ratios with the Galerkin-based bearing model. The reader is referred to section 5.1 for an elaborate discussion of this issue. It should, however, also be noted that for the same rotor-bearing systems, for which the solid journal bearings are described using the short bearing approximation, the same results were obtained for these bifurcation analyses as can be found in literature [152].

Porous journal bearing

Continuing with the symmetric rigid rotor-bearing systems mounted by porous journal bearings the same load cases as for the solid journal bearings are investigated for two different dimensionless permeabilities: $\Psi = 0.001$ and $\Psi = 0.002$.

Figure 6.6 shows the results for a porous journal bearing with $F_0^* = 1$ and $\Psi = 0.001$. For $a^* = 0$, the sub-critical Hopf bifurcation observed for $\Omega^* = 2.517$ is in agreement with the linear stability analysis. A branch of unstable periodic solutions emanates from the Hopf bifurcation which becomes stable at a fold bifurcation at $\Omega^* = 2.412$. In comparison to a solid journal bearing the periodic solution's maximum eccentricity ratio now increases more rapidly eventually exceeding $\varepsilon = 1$, corresponding to considerable elastic deformation of the rough surface. For porous bearings the load capacity does not approach infinity when $\varepsilon \to 1$, the asperity contact model also allows for $\varepsilon > 1$. However, prolonged operation in this regime might lead to local plastic deformation of the asperity contacts, which is why the results in these ranges should be interpreted with caution. The periodic solution remains stable at least till $\Omega^* = 20$ where the simulation was aborted.

An interesting case is encountered for all studied rotor-bearing systems containing porous bearings when $a^* = 0.1$. For this case, the usual period doubling bifurcations are encountered at $\Omega^* = 2.203$ and $\Omega^* = 3.035$. The periodicity of the solutions around these bifurcations is similar to that of the rotor-bearing systems containing solid journal bearings (see figure 6.4). Again, this is the case for all investigations for rotor-bearing systems with unbalance that contain porous journal bearings. The emanating branch of period-2 solutions from the first period doubling eventually coincides with another branch of period-2 solutions that emanates from the second period doubling bifurcation with a fold bifurcation at $\Omega^* = 7.125$. For this first branch, the solutions become stable at $\Omega^* = 2.176$. The second solution branch contains two additional period doubling bifurcations at $\Omega^* = 2.671$ and $\Omega^* = 2.705$, which are shown in detail in inset A. A branch of period-4 solutions which emanates from the first period doubling point is unstable and eventually coincides with the branch of period-2 solutions at the other period doubling point. An additional multiplier leaves the unit circle on the complex plane at a fold bifurcation for $\Omega^* = 2.601$, which is also shown in inset A. The periodicity of the solutions in inset A is indicated with circled numbers.

The cases with higher unbalance, $a^* = 0.2$ and $a^* = 0.3$, show a behavior very similar to the same load configurations for the solid journal bearings in figure 6.4. The locations of the period doubling bifurcations along the period-1 solution branch are almost the same as for the solid journal bearing. They are encountered at $\Omega^* = 2.04$ and $\Omega^* = 3.925$ for $a^* = 0.2$ and at $\Omega^* = 3.704$ and at $\Omega^* = 5.645$ for $a^* = 0.3$. This also is the case for the fold bifurcation for $a^* = 0.3$, which is found for $\Omega^* = 7.773$. However, for $a^* = 0.2$ the fold bifurcation now appears for a higher maximum eccentricity ratio at $\Omega^* = 12.97$. The load capacity for porous journal bearings is significantly smaller for high eccentricity ratios than for solid journal bearings and the maximum eccentricity ratio therefore increases and asperity contacts contribute more to the total load capacity.

The results for an increased load to $F_0^* = 2$ are shown in figure 6.7. Again, these results are comparable to the results shown in figure 6.5. The balanced rotor-bearing system has a super-critical Hopf bifurcation at $\Omega^* = 2.489$ which is in agreement with the linear stability analysis. From here a branch with periodic solutions emanates, which also becomes unstable between two fold bifurcations at $\Omega^* = 2.55$ and $\Omega^* = 2.514$. The fold bifurcations are found with higher maximum eccentricity ratio than for the solid journal bearing case and this again can be attributed to the decreased load capacity of porous journal bearings in this regime.

For $a^* = 0.1$, the two usual period doubling bifurcations are observed along the branch with period-1 solutions. Between these a branch of period-2 solutions is found. The initially unstable branch becomes stable between three sets of fold bifurcations: at $\Omega^* = 2.222, 6.195$, $\Omega^* = 3.0808, 3.081$ (inset B) and $\Omega^* = 2.65, 2.666$ and between a fold bifurcation at $\Omega^* = 2.578$ and a Neimark-Sacker point at $\Omega^* = 2.579$ (all of these four points are shown in detail in inset A). At the Neimark-Sacker point a pair of complex conjugated multipliers leaves the unit circle on the complex plane. One of these multipliers returns inside the unit circle at a fold bifurcation at $\Omega^* = 2.799$. In addition two period doubling bifurcations are found along the branch at $\Omega^* = 2.657, 2.605$, between which a branch of unstable period-4 solutions was detected. These points can also be found in inset A along with the periodicity of the solutions in circled numbers.

For the higher unbalances $a^* = 0.2$ and $a^* = 0.3$ the maximum eccentricity ratio increases even more in comparison to the analogue solid bearing cases. The period doubling bifurcations are found for similar angular frequencies. As expected all bifurcation points in the emanating branches appear at higher maximum eccentricity ratios.

Figures 6.8 and 6.9 show the results when the permeability is changed to $\Psi = 0.002$. These rotor-bearing systems show a very comparable behavior to those containing porous bearings with $\Psi = 0.001$. Because of this and because of reasons of brevity not all bifurcation points will be discussed again. The Hopf bifurcations for the balanced rotor-bearing systems again are found for the same angular frequencies as the onset frequency of instability resulting from the linear stability analysis. By raising the permeability the total load capacity for a given eccentricity ratio of the system is decreased which results in generally higher maximum eccentricity ratios for the fixed-point and periodic solutions. As a result the maximum eccentricity ratio also increases faster when the angular frequency is raised. New bifurcations are encountered for $a^* = 0.2$. For $F_0^* = 1$ the unstable branch emanating from the second period doubling point shows two additional fold bifurcations at $\Omega^* = 3.956$ and $\Omega^* = 3.831$, between which the periodic solutions briefly gains one unstable multiplier. An analogue pair of fold bifurcations appears for $F_0^* = 2$, but now the bifurcations occur closer to each other at $\Omega^* = 3.244$ and $\Omega^* = 3.301$. For this last case a new fold bifurcation shows up in the branch with period-2 solutions originating from the first period doubling bifurcation. Because of this the period doubling bifurcation is sub-critical instead of super-critical and the solutions in the emanating branch initially are unstable until the fold bifurcation is encountered.

The introduction of porous journal bearings does not significantly influence the onset of (local) linear stability in the hydrodynamical regime. However, for higher eccentricity ratios asperity contacts now play an important role and generate a significant part of the load capacity. Because of this the emanating branches with periodic solutions become stable at a higher eccentricity ratio for the balanced and moderately unbalanced ($a^* = 0.1$) rotor-bearing systems. Moreover, the eccentricity ratio can even exceed one as was observed for the balanced rotor systems, which might be an indication for running-in and pore closure. The transition from and to the regime with asperity contacts induces several additional bifurcation points for the moderately unbalanced ($a^* = 0.1$) rotor-bearing system, which complicate the dynamical behavior of porous journal bearings that are operated in the regime of mixed lubrication. The reader should however keep in mind that several additional effects play an important role in this regime and that the current results only give a first indication of how the rotor-bearing system could behave in this range. If the rotor-bearing systems with porous journal bearings are operated in the hydrodynamic regime they seem to be stable for higher rotor speeds than the systems containing solid journal bearings. When the load and/or permeability is increased such that asperity contacts become influential, the extent of the emanating branch with stable periodic solutions diminishes and the highest rotor speed of its stable solution decreases again.

6.2 Elementary rotor-bearing systems with flexible shafts

To introduce the influence of misalignment, the shaft is now taken to be flexible. Note that the system is still considered symmetric, such that only one set of DOF for the bearings has to be considered. For reasons of simplicity an Euler-Bernoulli beam element is used to model the elasticity of the shaft. Now that the rotor is considered flexible, its mass distribution becomes relevant and is defined using a shaft with a disk attached to it in its middle. See figure 6.10 for an illustration of the corresponding rotor-bearing system. The equations of motion have to be extended in order to include the extra degrees of freedom of the rotor as well as the tilting of the shaft inside the plain journal bearings. For reasons of simplicity the rotor-bearing system with a

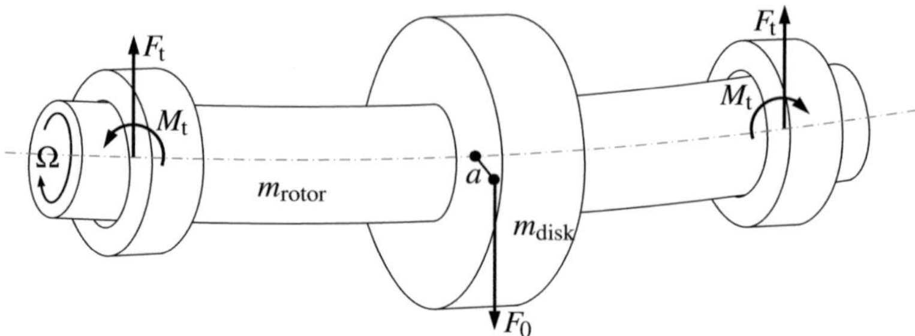

Figure 6.10: The rotor-bearing system in question with a flexible shaft, an unbalance a and load F_0.

flexible shaft is approximated using lumped masses. In doing so the shaft is divided lengthwise into four equal parts. The inertia of the outer most parts is concentrated at the bearings and the inertia of the other two parts is assumed to be concentrated at the disk. This leads to a model where only a mass $m_0 = m_{\text{disk}}/4$ is located at each bearing and a mass of $2(M_0 + 2m_0)$ (where $M_0 = m_{\text{rotor}}/4$) is located at the disk. The connecting elastic elements are assumed massless. To simplify matters even more the rotor is assumed to be perfectly balanced (i.e. $a = 0$). These simplifications and the resulting system are illustrated in figure 6.11. With these changes the equations of motion are given by

$$
\begin{cases}
\dfrac{1/2+\xi_m}{1+\xi_m}\varepsilon_x^{w\prime\prime} = F_x^{w*}/F_0^* \\[2mm]
\dfrac{1/2+\xi_m}{1+\xi_m}\varepsilon_y^{w\prime\prime} = F_y^{w*}/F_0^* - 1 \\[2mm]
\dfrac{1/2}{1+\xi_m}\varepsilon_x^{\prime\prime} = (2F_{t,x}^* - F_x^{w*})/F_0^* \\[2mm]
\dfrac{1/2}{1+\xi_m}\varepsilon_y^{\prime\prime} = (2F_{t,y}^* - F_y^{w*})/F_0^* \\[2mm]
\dfrac{1/4}{1+\xi_m}\phi_x^{\prime\prime} = g_1(M_{t,x}^* - M_x^{w*})/F_0^* - g_2\dfrac{1/4}{1+\xi_m}\Omega^*\phi_y^\prime \\[2mm]
\dfrac{1/4}{1+\xi_m}\phi_y^{\prime\prime} = g_1(M_{t,y}^* - M_y^{w*})/F_0^* + g_2\dfrac{1/4}{1+\xi_m}\Omega^*\phi_x^\prime
\end{cases}
\quad , \tag{6.6}
$$

where $\xi_m = m_{\text{disk}}/m_{\text{rotor}}$, the relative tilting angles and their dimensionless time derivatives between the journal and the bearing are indicated by $\phi_{x,y}$ and $\phi_{x,y}^\prime$ (see also figure 4.1) and $\varepsilon_{x,y}^w$ indicate the eccentricity ratios at the disk. The definition of F_0^* is analogue to the previous section and now includes the mass of the disk and the shaft. $F_{x,y}^{w*}$ and $M_{x,y}^{w*}$ are the reacting forces and moments of the Euler-Bernoulli beam element. These are given for a beam element with an applied load in the middle of the rotor and two moments on both axial ends (see figure 6.10 and 6.11). The deformation in the middle of the rotor and the tilting angle at the axial ends are resp. given by [61]

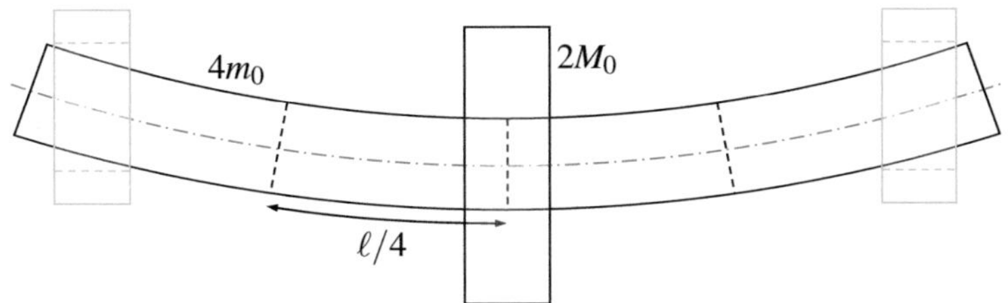

(a) Original rotor-bearing system

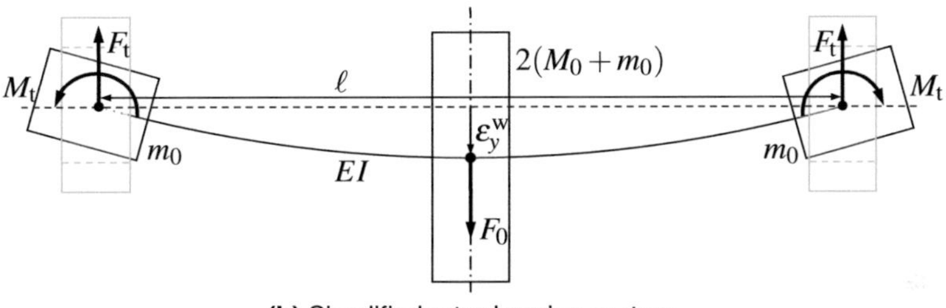

(b) Simplified rotor-bearing system

Figure 6.11: Schematic representations of the rotor-bearing system with a flexible shaft.

$$\delta_{x,y}^{\mathrm{w}} = \frac{1}{s}\left(F_{x,y}^{\mathrm{w}} + \frac{6}{\ell}M_{y,x}^{\mathrm{w}}\right) \quad \text{and} \quad \phi_{x,y}^{\mathrm{w}} = \frac{1}{s}\left(\frac{3}{\ell}F_{y,x}^{\mathrm{w}} + \frac{24}{\ell^2}M_{x,y}^{\mathrm{w}}\right), \tag{6.7}$$

where the shaft stiffness is $s = 48E^{\mathrm{w}}I^{\mathrm{w}}/\ell^3$ in which E^{w} refers to the elastic modulus of the shaft material, I^{w} to the area moment of inertia of the shaft and ℓ to the length of the shaft. Using these relationships the reacting rotor forces and moments become

$$F_{x,y}^{\mathrm{w}} = s(4\delta_{x,y}^{\mathrm{w}} - \ell\phi_{y,x}^{\mathrm{w}}) \quad \text{and} \quad M_{x,y}^{\mathrm{w}} = s\frac{\ell}{2}\left(\frac{\ell}{3}\phi_{x,y}^{\mathrm{w}} - \delta_{y,x}^{\mathrm{w}}\right) \tag{6.8}$$

or in dimensionless form

$$F_{x,y}^{\mathrm{w}*} = s^*\left(4(\varepsilon_{x,y} - \varepsilon_{x,y}^{\mathrm{w}}) - \beta\iota\phi_{y,x}^{\mathrm{w}}\right) \quad \text{and} \quad M_{x,y}^{\mathrm{w}*} = s^*\frac{\beta\iota}{6}\left(\beta\iota\phi_{x,y}^{\mathrm{w}} - 3(\varepsilon_{y,x} - \varepsilon_{y,x}^{\mathrm{w}})\right), \tag{6.9}$$

where $\beta = \ell/L$, $\iota = L/c$ and $s^* = (c^3/R_i^2)s/\eta\Omega_0 LD$ indicates the dimensionless stiffness. The parameters g_1 and g_2 are defined as

$$g_1 = \frac{12}{\iota}\frac{1}{3\kappa^2 + \beta^2/16} \quad \text{and} \quad g_2 = \frac{6\kappa^2}{3\kappa^2 + \beta^2/16}, \tag{6.10}$$

in which the moments of inertia of a cylinder with mass m_0, length $\ell/4$ and radius R are taken at both axial ends.

With this framework the following phase-space can be defined:

$$
\begin{cases}
u_1' = u_2 \\[4pt]
u_2' = \dfrac{1+\xi_{\mathrm{m}}}{1/2+\xi_{\mathrm{m}}}\, F_x^{\mathrm{w}*}/F_0^* \\[4pt]
u_3' = u_4 \\[4pt]
u_4' = \dfrac{1+\xi_{\mathrm{m}}}{1/2+\xi_{\mathrm{m}}}\,(F_y^{\mathrm{w}*}/F_0^* - 1) \\[4pt]
u_5' = u_6 \\[4pt]
u_6' = 2(1+\xi_{\mathrm{m}})(2F_{\mathrm{t},x}^* - F_x^{\mathrm{w}*})/F_0^* \\[4pt]
u_7' = u_8 \\[4pt]
u_8' = 2(1+\xi_{\mathrm{m}})(2F_{\mathrm{t},y}^* - F_y^{\mathrm{w}*})/F_0^* \\[4pt]
u_9' = u_{10} \\[4pt]
u_{10}' = 4g_1(1+\xi_{\mathrm{m}})(M_{\mathrm{t},x} - M_x^{\mathrm{w}*})/F_0^* - g_2\Omega^* u_{12} \\[4pt]
u_{11}' = u_{12} \\[4pt]
u_{12}' = 4g_1(1+\xi_{\mathrm{m}})(M_{\mathrm{t},y} - M_y^{\mathrm{w}*})/F_0^* + g_2\Omega^* u_{10}
\end{cases}
\qquad (6.11)
$$

where $u_1 = \varepsilon_x^{\mathrm{w}}$, $u_2 = \varepsilon_x^{\mathrm{w}\prime}$, $u_3 = \varepsilon_y^{\mathrm{w}}$, $u_4 = \varepsilon_y^{\mathrm{w}\prime}$, $u_5 = \varepsilon_x$, $u_6 = \varepsilon_x'$, $u_7 = \varepsilon_y$, $u_8 = \varepsilon_y'$, $u_9 = \phi_x$, $u_{10} = \phi_x'$, $u_{11} = \phi_y$ and $u_{12} = \phi_y'$.

6.2.1 Bifurcation analysis

The following features bifurcation diagrams for rotor-bearing systems containing a flexible shaft with and without the influence of misalignment in the porous bearings for two different stiffness parameters. The other parameters are chosen as $L/D = 1$, $F_0^* = 1$, $\xi_{\mathrm{m}} = 0.25$, $\iota = 400$, $\beta = 18$, $\Psi = 0.001$.

When misalignment is not taken into account the last four state variables of equation (6.11) are not considered. Moreover are the moments from equation (6.9) demanded to vanish for these rotor-bearing systems. If, additionally, all mass is assumed at the bearings (i.e. $\xi_{\mathrm{m}} \to \infty$) and the shaft stiffness approaches infinity, the model reduces to that of the previous section. Different in comparison to the results from the previous section is now the scalar measure for the bifurcation diagrams. The maximum eccentricity ratio now also depends on the axial

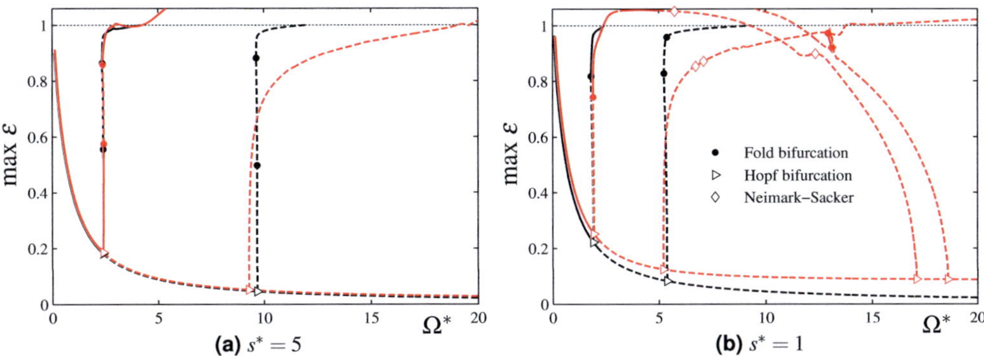

Figure 6.12: Bifurcation diagrams for balanced rotor-bearing systems that contain a flexible shaft for different shaft stiffness. The black curves indicate the results where bearing misalignment was not taken into account, whereas the red curves represent results for which misalignment in the bearings was taken into account.

coordinate when misalignment is considered and consequently the maximum of the eccentricity ratio at both axial ends of a bearing is taken as a scalar measure in the diagrams.

Figure 6.12 shows two bifurcation diagrams for relatively stiff rotors, namely $s^* = 1$ and $s^* = 5$. These feature results for which bearing misalignment was not taken into account (indicated with the black curves) and results for which bearing misalignment was accounted for (indicated with the red curves).

As is to be expected, the results with and without misalignment for the higher stiffness value $s^* = 5$ in figure 6.12a do not differ very much. For both rotor-bearing systems Hopf bifurcations are encountered around $\Omega^* = 2.37$ and $\Omega^* = 9.5$. The angular frequency of the first bifurcation has slightly diminished under the influence of the elastic shaft and its nature is now super-critical instead of sub-critical (see figure 6.6). The branches that emanate from the first Hopf bifurcation are very similar for both rotor-bearing systems. Even the locations of the two fold bifurcations that are encountered along the branch of periodic solutions are found for very similar parameters. A slight change is observed when the maximum eccentricity ratio approaches one. The maximum eccentricity ratio increases faster for the results that account for misalignment, which can be attributed to the fact that for misaligned bearings the amount of force which is generated through asperity contact is smaller than for aligned bearings. Moreover is the hydrodynamic load capacity of misaligned bearings smaller than for aligned bearings (see section 5.5). The additional moment due to misalignment does not seem to make a significant contribution in order to counteract this effect. As a result the maximum eccentricity ratio even exceeds one for misaligned bearings for a larger angular frequency range. When this is the case, the maximum eccentricity ratio actually is only exceeded at one or both axial ends of the bearings. In reality this would lead to severe edge loading, wear and pore closure, which significantly influences the bearing behavior. Such phenomena are not included in the current analysis and these results only give an indication of the behavior of the rotor-bearing system in this regime. Compared to the rigid rotor-bearing systems the second Hopf bifurcation is new, but is hard to interpret physically due to the unstable nature of its surrounding solution branches. Contrary to the first Hopf bifurcation, the maximum eccentricity of the solutions belonging to the branch that emanates from the second Hopf bifurcation increases more moderately when comparing the results with misalignment to those without misalignment. Two fold bifurcations between which the unstable periodic solution looses a stable multiplier are found along the emanating branch when misalignment is not considered, but do not appear when misalignment in the bearings is allowed.

The results for a rotor-bearing system with a more flexible rotor possessing the stiffness $s^* = 1$ are shown in figure 6.12b. Reducing the stiffness causes the Hopf bifurcation to appear at lower angular frequencies. For example, the first Hopf bifurcation is now encountered around $\Omega^* = 1.91$. The lower stiffness also has the effect that the differences between the results with and without misalignment become more apparent. The branch with fixed-point solutions now appears for larger maximum eccentricity ratios when misalignment is taken into account, whereas the eccentricity ratio has changed little for the rotor-bearing system with aligned bearings in comparison with the results from figure 6.12a. It can be reasoned that in rotor-bearing systems with misaligned bearings the smaller stiffness of the rotor results in higher tilting angles even for smaller eccentricity ratios. This in turn reduces the load capacity when compared to the aligned bearings. Although the effect is small it is still noticeable and can be explained using the results from figure 5.31. Lowering the stiffness also changed the nature of the Hopf bifurcation to sub-critical for both results with and without misalignment. The emanating branches with periodic solution are still very similar except when the maximum eccentricity approaches one. Here, the aligned bearings seem to perform worse than the misaligned bearings. The maximum eccentricity ratio of the latter result increases more gradually, which might be explained with the additional influence of carrying moment for misaligned bearings. Again, the maximum eccentricity ratio exceeds one when misalignment is accounted for in the bearings. Although in reality the results would be greatly influenced by edge loading, wear and pore closure in this range the branch is followed further nevertheless. It was found to coincide with another Hopf bifurcation at $\Omega^* = 17.1$ which was encountered along the branch with fixed-point solutions. This and another Hopf bifurcation (at $\Omega^* = 18.6$) were detected for the case with misalignment but not without misalignment. The branches that emanate from the second Hopf bifurcation around $\Omega^* = 5.2$ are still comparable to the results with $s^* = 5$. Now, however, several Neimark-Sacker points are encountered along the emanating branches for the case with misalignment, which also does not happen for the rotor-bearing systems with aligned bearings.

Concluding, it can be said that the introduction of misalignment in the bearings of a rotor-bearing system containing a flexible shaft does not have a significant influence on the essential dynamic behavior in the bearings of this system. When investigating the occurrence of bifurcation points and the shape of the solution branches several changes are observed, however, these do not contribute to the essential behavior of the rotor-bearing system. What is essential for the behavior of this system is the first Hopf bifurcation and its emanating branch of periodic solutions, which does not change significantly when misalignment in the bearings is taken into account. However, further research is necessary, since the current investigation only holds for balanced rotor-bearing systems where only two different stiffness values of the flexible shaft were studied. Inclusion of unbalance as well as a broader parameter study could still reveal an important influence of misalignment.

6.3 Multibody systems

To simulate multibody systems containing plain journal bearings the commercial software MSC.ADAMS is used. For this software a subroutine was created to calculate the forces and moments for a given set of (angular) displacements and (angular) velocities of the journal and the bearing (see figure 6.13). These are calculated using the proposed Galerkin-based bearing model. A brief overview of the architecture of the subroutine is given in this section. Subsequently, an example of a multibody system is given to which the approach was applied.

6.3.1 Inclusion in multibody systems

When considering a plain journal bearing in a multibody system, it is necessary to represent the angular velocities in the proper reference frame. Although ADAMS provides several functions to obtain the (angular) displacements and translational velocities in the proper reference frame, additional transformations are needed to obtain the correct values of $\dot{\phi}_x$ and $\dot{\phi}_y$. These are described in section 3.1.1.

The first time the subroutine is called several initializations are made. Aside from memory pre-allocations and parameter initializations several quantities are evaluated in advance to save

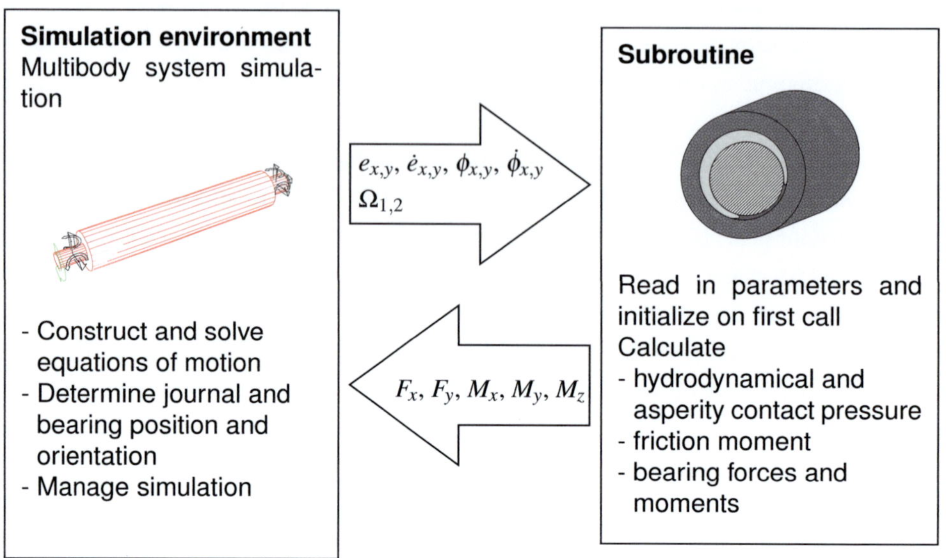

Figure 6.13: Subroutine implementation of the plain journal bearing for multibody system simulations.

calculation time. These include a one-dimensional lookup table for the forces arising from asperity contacts given by equation (4.11) and (4.14). Since no closed-from expressions exist for these integrals numerical integration is used to calculate these. The terms A_{nm} and B_{nm} appearing in the ansatz functions for the pressure and its time derivative (equations 4.28 and (4.29)) can also be evaluated prior to the actual calculations. The GNU Scientific Library (GSL) is used to evaluate the Bessel functions appearing in A_{nm} and B_{nm} (see appendix C) and to perform the numerical integration [53].

Every time the subroutine is called the systems of linear equations depicted in equation (4.31) and (4.33) have to be solved to obtain the coefficients in the ansatz function. For aligned plain journal bearings the left hand side of the system of linear equations is a symmetric positive definite band matrix and the Cholesky solver included in the Linear Algebra PACKage (LAPACK) is used to efficiently solve these systems [3]. The LAPACK also includes the Basic Linear Algebra Subprograms (BLAS) [91]. For misaligned plain journal bearings the system of linear equations contains several additional non-zero diagonals. To efficiently solve these systems CHOLMOD is used [35], which also makes use of the BLAS and the LAPACK.

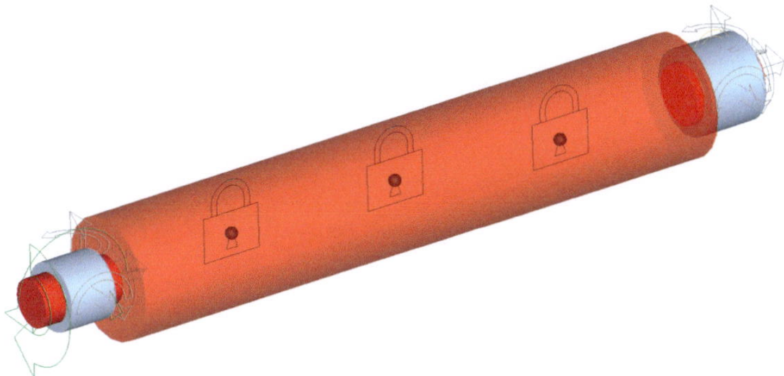

Figure 6.14: ADAMS model of an armature mounted by two porous journal bearings

6.3.2 Practical application

As a practical application the proposed bearing model was used to simulate the dynamical behavior of an armature mounted by two porous journal bearings. To approximate the armature's moment of inertia several unbalances have been attached to a rotor to represent its essential behavior. The influence of the housing's deformation surrounding the left hand side bearing on the dynamical behavior of the rotor's center in the right hand side bearing was studied using the multibody simulation software MSC.ADAMS.

The housing's deformation gives rise to tilting and displacements of the bush causing misalignment and translational motion in the porous journal bearing on the left hand side. This in turn affects the rotor's motion and therefore also influences the dynamical behavior of the armature. Too large eccentricity ratios can be problematic for the motor surrounding the armature and finding their limits and the phenomena that cause these were key aspects of the investigation. The only way to investigate these effects is by observing the motion in the right hand side bearing. The orbits in the right hand side bearing for several rotor speeds, tilting angles and positions change for the left hand side bearing were compared to orbits measured in the right hand side bearing.

Chapter 7

Conclusion & Outlook

7.1 Conclusion

Plain journal bearings find a wide range of applications for mounting rotating machinery. Especially the porous journal bearing gained in popularity during the previous century due to its advantageous lubrication properties and cost efficiency. At the same time, increasing operating condition requirements and cost reduction efforts have widened the load and rotor speed range for porous journal bearings. Along with the rising need to simulate the dynamic behavior of entire products this demands for fast and robust models for plain journal bearings.

Plain journal bearing model

Most models for porous journal bearing are based on advanced numerical discretization schemes. Although these methods can be very accurate, practical applications often are numerically very costly, time-consuming and therefore not suitable for many applications like multibody simulations for instance. Simpler analytical approximations on the other hand fail to describe the essential physical behavior.

The porous journal bearing problem is governed by the Reynolds equation for the pressure in the fluid film and Darcy's law for the fluid pressure in the porous bush. To solve the Reynolds equation Galerkin's method is used to obtain a low-dimensional semi-analytical approximation of the pressure function. Advantages of this approach are that only a relatively small, sparse, band-structured, symmetric, positive definite system of linear equations has to be solved. By including the solution of Darcy's law for the porous medium a priori into the pressure ansatz function, the method is used to obtain a compact model for the multiphysical problem of porous journal bearings without any additional numerical costs. Freely available linear algebra packages allow for efficient solution schemes of the resulting systems of linear equations. Due to the semi-analytical approach unsteady-state stiffness and damping coefficients (i.e. these coefficients also are valid when the journal bearing is not in an equilibrium position) can be calculated by symbolic derivation instead of applying additional numerical differentiation schemes. The method can be modified to account for more complex phenomena such as the influence of journal misalignment or surface roughness approximated with flow factors. The additional influence of asperity contacts is included by adopting a contact model that approximates the rough surfaces as a distribution of paraboloids whose contact pressure is calculated using Hertzian theory.

It is shown that the proposed bearing model is able to approximate the solution more efficiently than using a numerical discretization scheme. A small number of ansatz functions suffices in the majority of the cases to accurately approximate the pressure function. Only for solid journal bearings with very small fluid film thicknesses the necessary number of ansatz function increases rapidly, which, however, is not an issue for porous journal bearings.

Verification and validation of the proposed bearing model

Comparisons of impedance maps of the proposed bearing model (which uses the GÜMBEL cavitation condition) with a more accurate numerical discretization method (that employs the more accurate REYNOLDS cavitation condition) show that for typical bearing geometries the proposed model approximates the latter solution very well for solid as well as porous journal bearings.

Experimental results of the friction coefficient in the hydrodynamic regime of sintered bronze bearings correspond well to the friction calculated with the porous journal bearing model for low loads, whereas for high loads the solid journal bearing model appears to agree better with the experimental results. This can be explained by the phenomenon of running-in, which is not incorporated into the present model. For high loads this causes pore closure, which significantly lowers the permeability and makes the bearing behavior more like that of a solid journal bearing. The proposed model is able to qualitatively predict the trend of mixed lubrication. Unfortunately, the experimental results were inconclusive in this range and consequently the proposed bearing model could not be validated here. Comparisons with the sintered iron bearings were unsuccessful, due to the fact that these bearings ran for the majority of the time under mixed lubrication conditions. The absence of usable roughness parameters and the fact that the proposed bearing model is not able to simulate the effect of running-in and pore closure made these comparisons problematic.

It has been shown that the influence of elastohydrodynamic lubrication only becomes influential for relatively small radial clearances, however, this effect can still be considered minor under normal operating conditions. For larger radial clearances the bearing's elastic deformation is practically insignificant, even for more extreme operating conditions.

Steady state plain journal bearing characteristics

Steady-state stiffness and damping coefficients, the load capacity and the equilibrium positions of the proposed model for the short and long bearings approximations were successfully compared with analytical methods.

Subsequently, the influence of the added Darcy's law showed that porous metal bearings are more likely to run at higher eccentricity ratios. This lead to the need for a model which incorporates the influence of rough surfaces through flow factors on the hydrodynamical pressure and through asperity contacts, which are more likely at small film thicknesses.

The influence of surface roughness on the hydrodynamical pressure generally increases the load capacity and stiffness coefficients at higher eccentricity ratios ($\varepsilon \approx 0.5 \ldots 0.9$) of a porous journal bearing, but does not seem to have a significant effect on the damping properties. For very high eccentricity ratios ($\varepsilon > 0.9$) the influence of surface roughness seems to be reversed and the load capacity and stiffness coefficients decrease with respect to a smooth porous journal bearing. This has to do with the fact that the used flow factors have a large uncertainty for small fluid film thicknesses. This matter of fact combined with the assessment that in the face of asperity contacts the influence of flow factors is relatively small, lead to the decision to exclude the flow factors from the subsequent analyses.

The implications of misalignment on the load capacity are less significant for large and moderate film heights. When the film thickness function becomes relatively small the load capacity of misaligned bearings can be significantly lower than that of aligned bearings. However, under normal operating conditions and moderate film thickness misalignment will most likely become influential through the additional moment capacity.

Dynamics of elementary rotor-bearing systems containing plain journal bearings

The dynamical behavior of rotor-bearing systems containing solid and porous journal bearings was studied. To this end the unsteady-state stiffness and damping coefficients were used to give the Jacobian in explicit form, which was used to improve robustness and performance of the numerical bifurcation branch continuation analysis and to ascertain the onset frequency of linear instability.

The onset rotor speed of (local) instability is lower for a rotor-bearing system containing porous journal bearings than for a system with solid journal bearings. Increasing the permeability of the porous journal bearings decreases the onset rotor speed of instability. While balanced rotor-bearing systems containing solid journal bearings are always stable beyond a certain eccentricity ratio in the hydrodynamic regime, the same systems containing porous bearings only exceed this limit in the range of mixed lubrication.

To study the dynamical behavior of a rotor-bearing system beyond the onset frequency of instability so-called path-following software was used. The solutions of the rotor-bearing systems were followed for changing rotor-speed and for several applied loads, unbalances and permeabilities. The investigations showed that in most cases a higher rotor unbalance decreases the onset frequency of (local) instability, but that this is – depending on the applied load – not always the case. In general it can be said that for higher loads the branch of stable solutions which emanates from the first Hopf or period doubling bifurcation loses its stability for lower rotor speeds.

For rotor-bearing systems containing porous bearings the decreased load capacity at high eccentricity ratios induces asperity contacts, which complicate the dynamical behavior in this regime. Even for balanced rotor-bearing systems the maximum eccentricity ratio can now exceed one, which is a strong indication for the occurrence of running-in and pore closure. When the rotor-bearing systems with porous journal bearings are operated in the hydrodynamic regime the emanating branch of periodic solutions from the first Hopf or period doubling bifurcation seems to be stable for higher rotor speeds than the systems containing solid journal bearings. When the load and/or permeability is increased such that asperity contacts become influential, the extent of the emanating branch with stable periodic solutions diminishes and the highest rotor speed of its stable solutions decreases in comparison to the systems with solid journal bearings.

Preliminary results of a balanced rotor-bearing system with a flexible rotor show that the influence of bearing misalignment does not significantly affect its essential dynamical behavior. An extended investigation which also accounts for unbalanced rotors and a broader parameter range still might reveal an influence though.

An example was presented to which the proposed bearing model was successfully applied to simulate the dynamical behavior of a multibody system in the form of an armature mounted by two porous journal bearings.

Concluding, the proposed Galerkin-based bearing model is a fast and efficient approach to study the dynamical behavior of (porous) journal bearings including complex phenomena such as the influence of surface roughness on the hydrodynamical pressure, asperity contacts and journal misalignment. The method was verified by comparison with other numerical methods and experimentally validated. With this it has been shown that the proposed bearing model is a viable approach to study the dynamical behavior of (multibody) systems containing plain journal bearings, even beyond the onset of (local) instability.

7.2 Outlook

In this last section an overview is given of potential methods of improvement, extensions to account for certain phenomena and issues which could be investigated in future work.

- The measurements that were employed for the porous journal bearings, which were used to validate the proposed bearing model should be repeated and extended. Firstly, to accurately estimate the error of measurement and secondly to obtain acceptable assessments of the parameters related to the roughness profile and friction properties.

- In relationship to the previous point, experiments could be undertaken depicting the dynamical properties of porous journal bearings. Publications of such measurements are not known to the author. Initially, these could be in the form of steady-state stiffness and damping measurements and could be extended to unsteady-state properties in a later stadium.

- Depending on the applied load, rotor speed and unbalance, understanding the regime of mixed lubrication in porous journal bearings can be of great importance. A modeling approach that describes the running-in process and its combined influence on the porous bush's surface roughness properties and permeability distribution will be an important step in this process. The measurements in this work showed that the applied load has a major contribution to this effect and additional measurements investigating this should be undertaken to validate the previously proposed modeling approach.

- The effect of running-in, misalignment but also differing bearing geometries in porous journal bearings induce an anisotropic and/or non-homogeneous permeability distribution. This topic has not been thoroughly studied in literature and could be investigated in future work.

- In this work it was shown that the phenomenon of elastohydrodynamic lubrication does not need to be considered under normal operating conditions in porous journal bearings. However, under extreme loads or for extreme misaligned bearings this effect should be taken in account.

- Almost all literature on porous journal bearings assumes the temperature distribution to be homogeneous, with the exception of [17]. The temperature has, however, an important influence on the fluid viscosity and therefore also on the bearing's behavior, which justifies such an investigation.

- Regarding the dynamical behavior of plain journal bearings, the flow factors as well as the asperity contact description utilized in this work only depend on the fluid film thickness function. No studies are known where the dependence of these phenomena on the film thickness function's time derivative is taken into account. New descriptions for the flow factors as well as the asperity contacts could be developed that account for this.

- Accurately predicting the pressure profile in the fluid film is currently still problematic. Several studies in literature have shown promising results using advanced cavitation algorithms [103, 150]. However, many questions remain unanswered concerning the interaction of cavitation in the fluid film and fluid exchange with the porous medium and research on this topic should be continued.

- Although the current investigation on the influence of bearing misalignment has shown that this does not significantly affect the dynamical behavior of the concurring flexible rotor-bearing system, further investigations still are in order. In these investigations rotor unbalance could be taken into account and a broader parameter study could be considered.

Appendix A

Derivation of the Reynolds equation

This appendix describes the derivation of the Reynolds equation under the assumptions that were introduced in chapter 2. Of course, in general the following derivation does not hold, which is for example the case when the viscosity and/or density are assumed to be pressure and/or temperature dependent. Corresponding derivations can be found e.g. in [28]. Nevertheless, the current derivation is found in many instances in literature, e.g. [21] and it is given here to explain the coupling with Darcy's law and the special velocity slip conditions that occur in porous journal bearings. Referring to figure 2.1 and making the same assumptions postulated

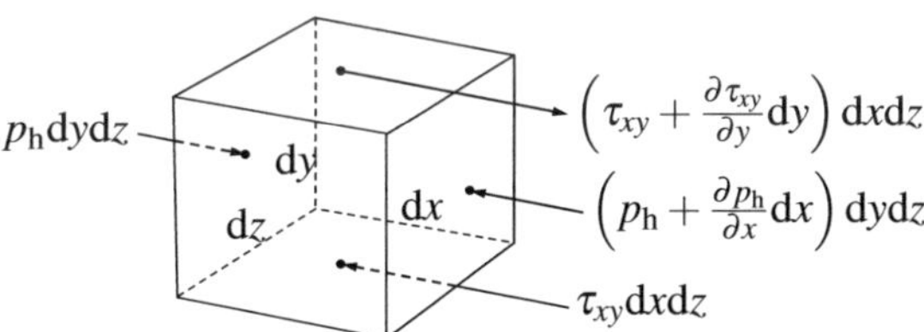

Figure A.1: An infinitesimal volume element with force balance in the x-direction.

in chapter 2 we consider an infinitesimal volume element with base lengths $\mathrm{d}x$, $\mathrm{d}y$ and $\mathrm{d}z$. The force balance in the x-direction is illustrated in figure A.1. Due to equilibrium conditions the force acting to the right must be canceled out by the forces acting to the left (in this case the shear stress τ_{xy} and the pressure p_h):

$$p_\mathrm{h}\mathrm{d}y\mathrm{d}z + \left(\tau_{xy} + \frac{\partial \tau_{xy}}{\partial y}\mathrm{d}y\right)\mathrm{d}x\mathrm{d}z = \tau_{xy}\mathrm{d}x\mathrm{d}z + \left(p_\mathrm{h} + \frac{\partial p_\mathrm{h}}{\partial x}\mathrm{d}x\right)\mathrm{d}y\mathrm{d}z. \tag{A.1}$$

A completely analogue equality can be evaluated for the z-direction. These relationships eventually lead to

$$\frac{\partial p_h}{\partial x} = \frac{\partial \tau_{xy}}{\partial y} \quad \text{and} \quad \frac{\partial p_h}{\partial z} = \frac{\partial \tau_{zy}}{\partial y}. \tag{A.2}$$

As was stated in chapter 2 the fluid is assumed to behave like a Newtonian fluid. Therefore, the shear stresses in x- and z-direction are respectively governed by

$$\tau_{xy} = \eta \frac{\partial u}{\partial y} \quad \text{and} \quad \tau_{zy} = \eta \frac{\partial w}{\partial y}, \tag{A.3}$$

where u and w are the velocities of a fluid particle in x- and z-direction respectively. Substituting the equations from (A.3) into (A.2) gives

$$\frac{\partial^2 u}{\partial y^2} = \frac{1}{\eta} \frac{\partial p_h}{\partial x} \quad \text{and} \quad \frac{\partial^2 w}{\partial y^2} = \frac{1}{\eta} \frac{\partial p_h}{\partial z}. \tag{A.4}$$

One can arrive at the same result by simplifying the Navier-Stokes equations using the assumptions stated at the beginning of chapter 2. After integration and applying the boundary conditions from figure 2.1 u and w are obtained

$$u = \frac{1}{2\eta} \frac{\partial p_h}{\partial x} y(y - H) + (U_2 - U_1)\frac{y}{H} + U_1, \tag{A.5}$$

$$w = \frac{1}{2\eta} \frac{\partial p_h}{\partial z} y(y - H) + (W_2 - W_1)\frac{y}{H} + W_1. \tag{A.6}$$

By integrating the fluid velocities over the film thickness H the rates of flow per unit width in x- and z- direction are obtained:

$$q_x = \int_0^H u\,dy = \frac{U_1 + U_2}{2}H - \frac{H^3}{12\eta}\frac{\partial p_h}{\partial x}, \tag{A.7}$$

$$q_z = \int_0^H w\,dy = \frac{W_1 + W_2}{2}H - \frac{H^3}{12\eta}\frac{\partial p_h}{\partial z}. \tag{A.8}$$

Now consider a volume element or column which extends over the complete film thickness H, shown in figure A.2. Aside from the rate of flow in the x- and z-direction the volume of the element is influenced by the velocity terms V_1 and V_2 which correspond to the velocities of the top ($y = H$) and bottom ($y = 0$) boundaries of the column. Normally, these terms refer to the bottom and top surface velocity, but when fluid particles enter or leave the volume element an additional velocity can be included in these terms to account for this. The latter is applied in this thesis to model the flow between the porous bush and the fluid film in porous journal bear-

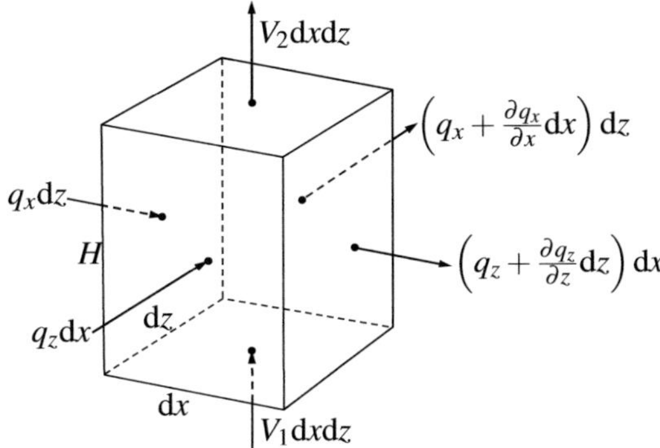

Figure A.2: Flow in a volume element which extends over the complete film thickness.

ings and is included in V_1, which now represents the fluid velocity plus the velocity belonging to the top surface. The other velocity V_2 solely represents the change in film thickness over time due to the motion of the wall, which is needed to include the dynamical behavior in plain journal bearings.

When considering an infinitesimal volume element inside the element stretching over the complete film thickness with height dy, the flow should abide the relationship

$$\frac{\partial u}{\partial x} + \frac{\partial v}{\partial y} + \frac{\partial w}{\partial z} = 0 \tag{A.9}$$

and when integrating this expression over the complete film height it again should equal zero

$$\int_0^H \left(\frac{\partial u}{\partial x} + \frac{\partial v}{\partial y} + \frac{\partial w}{\partial z}\right) dy = \frac{\partial q_x}{\partial x} - U_2 \frac{\partial H}{\partial x} + \frac{\partial q_z}{\partial z} - W_2 \frac{\partial H}{\partial Z} + (V_2 - V_1) = 0. \tag{A.10}$$

Finally, substituting the expressions for q_x and q_z into this equation leads to the Reynolds equation:

$$\frac{\partial}{\partial x}\left(\frac{H^3}{12\eta}\frac{\partial p_{\mathrm{h}}}{\partial x}\right) + \frac{\partial}{\partial z}\left(\frac{H^3}{12\eta}\frac{\partial p_{\mathrm{h}}}{\partial z}\right) = H\frac{\partial}{\partial x}\frac{U_1 + U_2}{2} + \frac{U_1 - U_2}{2}\frac{\partial H}{\partial x} \cdots$$
$$\cdots + H\frac{\partial}{\partial z}\frac{W_1 + W_2}{2} + \frac{W_1 - W_2}{2}\frac{\partial H}{\partial z} + (V_2 - V_1). \tag{A.11}$$

Appendix B

Parameters in the Reynolds equation

This appendix lists the parameters appearing in equation (4.2), (4.18) and (4.21). The parameters are given as functions of the journal position (ε_x, ε_y) and translational velocity (ε_x', ε_y') at the mid-plane, the rotational velocity (Ω^*) and the misalignment angles (ϕ_x, ϕ_y) and their derivatives with respect to time (ϕ_x', ϕ_y') at the mid-plane as well as the radial clearance to bearing length ratio $\iota = c/L$.

The eccentricity ratio is given by

$$\varepsilon = \sqrt{\varepsilon_x^2 + \varepsilon_y^2}. \tag{B.1}$$

The attitude angle is defined by

$$\sin\gamma = \varepsilon_x/\varepsilon \quad \text{and} \quad \cos\gamma = -\varepsilon_y/\varepsilon. \tag{B.2}$$

The norm of the pure squeeze velocity vector is

$$v_s^* = \sqrt{\varepsilon'^2 + \left(\varepsilon(\gamma' - \tfrac{1}{2}\Omega^*)\right)^2}, \tag{B.3}$$

where

$$\varepsilon' = \frac{\varepsilon_x \varepsilon_x' + \varepsilon_y \varepsilon_y'}{\varepsilon} \quad \text{and} \quad \gamma' = \frac{\varepsilon_x \varepsilon_y' - \varepsilon_y \varepsilon_x'}{\varepsilon^2}.$$

The angle between the eccentricity vector and the pure squeeze velocity vector is defined by

$$\sin\alpha = -\frac{\varepsilon(\gamma' - \tfrac{1}{2}\Omega^*)}{v_s^*} \quad \text{and} \quad \cos\alpha = \frac{\varepsilon'}{v_s^*}. \tag{B.4}$$

The length of the journal axis projection at the mid-plane and its time derivative are

$$\bar{\varepsilon} = \iota\sqrt{\phi_x^2 + \phi_y^2} \quad \text{and} \quad \bar{\varepsilon}' = \iota^2\frac{\phi_x\phi_x' + \phi_y\phi_y'}{\bar{\varepsilon}}. \tag{B.5}$$

Finally, the angle between the journal axis projection at the mid-plane and the eccentricity vector is defined with

$$\sin\psi = -\iota\frac{\varepsilon_x\phi_x + \varepsilon_y\phi_y}{\varepsilon\,\bar{\varepsilon}} \quad \text{and} \quad \cos\psi = +\iota\frac{\varepsilon_x\phi_y - \varepsilon_y\phi_x}{\varepsilon\,\bar{\varepsilon}} \tag{B.6}$$

and its time derivative becomes therewith

$$\psi' = \iota\frac{\phi_x\phi_y' - \phi_y\phi_x'}{\bar{\varepsilon}} - \gamma'. \tag{B.7}$$

Appendix C

Solution of the Laplace equation

This appendix lists the solution coefficients in the solution of the Laplace equation depending on which boundary condition is used. Referring to equation (4.28) and equation (4.29) these are the terms A_{nm} and B_{nm}. For both solutions $\kappa_o = R_o/L$, I_n and K_n are the modified Bessel functions of the first and second kind, respectively, and

$$\beta_m = \begin{cases} m\pi & \text{with misalignment} \\ \pi(2m-1) & \text{without misalignment} \end{cases}.$$

C.1 Dirichlet boundary condition

For the Dirichlet boundary condition $(\widehat{p}_h^*(\theta,z^*,r^*=R_o/R_i)=0)$ the coefficients in equation (4.28) are

$$A_{nm} = I_n(\beta_m\kappa) - \frac{I_n(\beta_m\kappa_o)}{K_n(\beta_m\kappa_o)} K_n(\beta_m\kappa) \tag{C.1}$$

and the coefficients in equation (4.29) are

$$B_{nm} = \tfrac{1}{2}\kappa\beta_m\left(I_{n-1}(\beta_m\kappa) + I_{n+1}(\beta_m\kappa) + \frac{I_n(\beta_m\kappa_o)}{K_n(\beta_m\kappa_o)}\left(K_{n-1}(\beta_m\kappa) + K_{n+1}(\beta_m\kappa)\right)\right). \tag{C.2}$$

C.2 Neumann boundary condition

For the Neumann boundary condition $(\partial \widehat{p}_{\mathrm{h}}^*(\theta, z^*, r^* = R_o/R_i)/\partial r^* = 0)$ the coefficients in equation (4.28) are

$$A_{nm} = I_n(\beta_m \kappa) + \frac{I_{n-1}(\beta_m \kappa_o) + I_{n+1}(\beta_m \kappa_o)}{K_{n-1}(\beta_m \kappa_o) + K_{n+1}(\beta_m \kappa_o)} K_n(\beta_m \kappa) \tag{C.3}$$

and the coefficients in equation (4.29) are

$$B_{nm} = \tfrac{1}{2}\kappa \beta_m \left(I_{n-1}(\beta_m \kappa) + I_{n+1}(\beta_m \kappa) - \frac{I_{n-1}(\beta_m \kappa_o) + I_{n+1}(\beta_m \kappa_o)}{K_{n-1}(\beta_m \kappa_o) + K_{n+1}(\beta_m \kappa_o)} \left(K_{n-1}(\beta_m \kappa) + K_{n+1}(\beta_m \kappa) \right) \right). \tag{C.4}$$

Appendix D

Flow factors

The flow factors from [125] are given by

$$
\phi_{\theta,z}^{f} = \begin{cases} 1 - Ce^{-\tilde{r}H/\sigma_{\mathrm{h}}} & \text{for } \tilde{\gamma} \leq 1 \\ 1 + C(H/\sigma_{\mathrm{h}})^{-\tilde{r}} & \text{for } \tilde{\gamma} > 1 \end{cases},
\tag{D.1}
$$

$$
\phi_{s}^{f} = \left(\frac{\sigma_{\mathrm{h},1}}{\sigma_{\mathrm{h}}}\right)^{2} \Phi_{s}(H/\sigma_{\mathrm{h}}, \tilde{\gamma}_{1}) - \left(\frac{\sigma_{\mathrm{h},2}}{\sigma_{\mathrm{h}}}\right)^{2} \Phi_{s}(H/\sigma_{\mathrm{h}}, \tilde{\gamma}_{2}),
\tag{D.2}
$$

where

$$
\Phi_{s} = \begin{cases} A_{1}(H/\sigma_{\mathrm{h}})^{\alpha_{1}} e^{-\alpha_{2}H/\sigma_{\mathrm{h}} + \alpha_{3}(H/\sigma_{\mathrm{h}})^{2}} & \text{for } H/\sigma_{\mathrm{h}} \leq 5 \\ A_{2}e^{-0.25H/\sigma_{\mathrm{h}}} & \text{for } H/\sigma_{\mathrm{h}} > 5 \end{cases}.
\tag{D.3}
$$

The constants C, r, A_1, A_2, $\alpha_{1,2,3}$ are given as functions of the Peklenik numbers $\tilde{\gamma}$ in table D.1.

$\tilde{\gamma}$	C	$\tilde{r}$	A_1	α_1	α_2	α_3	A_2	range
1/6	1.38	0.42	1.962	1.08	0.77	0.03	1.754	$H/\sigma_{\mathrm{h}} > 1$
1	0.90	0.56	1.899	0.98	0.92	0.05	1.126	$H/\sigma_{\mathrm{h}} > 0.5$
6	0.52	1.50	1.290	0.62	1.09	0.08	0.388	$H/\sigma_{\mathrm{h}} > 0.5$

Table D.1: Coefficients from [125] in equation (D.1) and equation (D.3).

References

[1] ADILETTA, G., GUIDO, AR, & ROSSI, C. 1996. Chaotic motions of a rigid rotor in short journal bearings. *Nonlinear Dynamics*, **10**(3), 251–269.

[2] ALMQVIST, A., & DASHT, J. 2006. The homogenization process of the Reynolds equation describing compressible liquid flow. *Tribology international*, **39**(9), 994–1002.

[3] ANDERSON, E., BAI, Z., BISCHOF, C., BLACKFORD, S., DEMMEL, J., DONGARRA, J., DU CROZ, J., GREENBAUM, A., HAMMARLING, S., MCKENNEY, A., & SORENSEN, D. 1999. *LAPACK Users' Guide*. Third edn. Philadelphia, PA: Society for Industrial and Applied Mathematics.

[4] BARRETT, L.E., ALLAIRE, P.E., & GUNTER, E.J. 1979. A finite length bearing correction factor for short bearing theory. *In: American Society of Mechanical Engineers and American Society of Lubrication Engineers, Lubrication Conference, Dayton, Ohio*.

[5] BARTEL, D. 2001. *Berechnung von Festkörper- und Mischreibung bei Metallpaarungen*. Ph.D. thesis, Universität Magdeburg.

[6] BARTZ, W.J., *et al.* . 1993. *Selbstschmierende und wartungsfreie Gleitlager*. Expert-Verlag.

[7] BASTANI, YASER, & DE QUEIROZ, MARCIO. 2010. A New Analytic Approximation for the Hydrodynamic Forces in Finite-Length Journal Bearings. *Journal of Tribology*, **132**(1), 014502.

[8] BEAR, J. 1988. *Dynamics of fluids in porous media*. Dover publications.

[9] BEAVERS, G.S., & JOSEPH, D.D. 1967. Boundary conditions at a naturally permeable wall. *Journal of Fluid Mechanics Digital Archive*, **30**(01), 197–207.

[10] BHUSHAN, B. 1998. Contact mechanics of rough surfaces in tribology: multiple asperity contact. *Tribology Letters*, **4**(1), 1–35.

[11] BHUSHAN, B. 2000. *Tribology: Friction, wear, and lubrication.* Vol. 210. CRC Press, Boca Raton.

[12] BHUSHAN, B. 2002. *Introduction to tribology.* Wiley.

[13] BOBACH, L. 2008. *Simulation dynamisch belasteter Radialgleitlager unter Mischreibungsbedingungen.* Ph.D. thesis, Universität Magdeburg.

[14] BOBACH, L., BARTEL, D., & DETERS, L. 2007. Das dynamisch belastete Radialgleitlager unter dem Einfluss elastischer Verformungen der Lagerumgebung. *Tribologie und Schmierungstechnik*, **54**(1), 5–13.

[15] BOEDO, S., & BOOKER, J. F. 2004. Classical Bearing Misalignment and Edge Loading: A Numerical Study of Limiting Cases. *Journal of Tribology*, **126**(3), 535–541.

[16] BOOKER, J.F. 1965. Dynamically loaded journal bearings: mobility method of solution. *ASME Journal of Basic Engineering*, **87**, 537–546.

[17] BOUBENDIR, S., LARBI, S., & BENNACER, R. 2011. Numerical study of the thermo-hydrodynamic lubrication phenomena in porous journal bearings. *Tribology International*, **44**(1), 1 – 8.

[18] BRAUN, AL. 1982. Porous bearings. *Tribology International*, **15**(5), 235–242.

[19] BRAUN, M.J., & HANNON, W.M. 2010. Cavitation formation and modeling: a review. *Proceedings of the Institution of Mechanical Engineers, Part J: Journal of Engineering Tribology*, **224**(9), 839–863.

[20] BRINKMAN, H.C. 1949. A calculation of the viscous force exerted by a flowing fluid on a dense swarm of particles. *Applied Scientific Research*, **1**(1), 27–34.

[21] CAMERON, A. 1966. *The Principles of Lubrication.* Longmans.

[22] CAMERON, A., MORGAN, V.T., & STAINSBY, A.E. 1962. Critical Conditions for Hydrodynamic Lubrication of Porous Metal Bearings. *Proceedings of the Institution of Mechanical Engineers, Part J: Journal of Engineering Tribology*, **176**(28), 761–769.

[23] CAPONE, E. 1970. Lubrication of axially undefined porous bearings. *Wear*, **15**, 157–170.

[24] CAPONE, E., & D'AGOSTINO, V. 1974. Oil whirl in porous metal bearings. *Meccanica,* **9**(2), 121–129.

[25] CHILDS, D., MOES, H., & VAN LEEUWEN, H. 1977. Journal Bearing Impedance Descriptions For Rotordynamic Applications. *Journal of Lubrication Technology,* **99**, 198–214.

[26] CHRISTENSEN, H. 1969. Stochastic models for hydrodynamic lubrication of rough surfaces. *Proceedings of the Institution of Mechanical Engineers,* **184**(1969), 1013–1026.

[27] CHRISTENSEN, H., & TONDER, K. 1971. The hydrodynamic lubrication of rough bearing surfaces of finite width. *ASME Journal of Lubrication Technology,* **93**(3), 324–330.

[28] CIMATTI, G. 1983. How the Reynolds equation is related to the Stokes equations. *Applied Mathematics and Optimization,* **10**(1), 267–274.

[29] CONRY, T.F., & CUSANO, C. 1974. On the stability of porous journal bearings. *ASME J Eng Ind,* **96**(2), 585–590.

[30] CROOIJMANS, M.T.M., BROUWERS, H.J.H., CAMPEN, D.H., & DE KRAKER, A. 1990. Limit cycle predictions of a nonlinear journal-bearing system. *Journal of engineering for industry,* **112**, 168–171.

[31] CUSANO, C. 1972. Lubrication of Porous Journal Bearings. *Journal of Lubrication Technology,* **94**(1), 69–73.

[32] D'AGOSTINO, V., RUGGIERO, A., & SENATORE, A. 2006. Approximate model for unsteady finite porous journal bearings fluid film force calculation. *Proceedings of the Institution of Mechanical Engineers, Part J: Journal of Engineering Tribology,* **220**(3), 227–234.

[33] D'AGOSTINO, V., RUGGIERO, A., & SENATORE, A. 2009. Unsteady oil film forces in porous bearings: analysis of permeability effect on the rotor linear stability. *Meccanica,* **44**(2), 207–214.

[34] DARCY, H. 1856. Les fontaines publiques de la ville de Dijon. *Dalmont, Paris,* **647**.

[35] DAVIS, T.A., & HAGER, W.W. 2009. Dynamic supernodes in sparse Cholesky update/downdate and triangular solves. *ACM Transactions on Mathematical Software (TOMS),* **35**(4), 1–23.

[36] DE KRAKER, A., VAN OSTAYEN, R.A.J., & RIXEN, D.J. 2007. Calculation of Stribeck curves for (water) lubricated journal bearings. *Tribology International,* **40**(3), 459–469.

[37] DE KRAKER, A., VAN OSTAYEN, R.A.J., & RIXEN, D.J. 2010. Development of a texture averaged Reynolds equation. *Tribology International*, **43**, 2100–2109.

[38] DE KRAKER, B. 2009. *Introductory Rotordynamics*. Shaker Verlag.

[39] DHOOGE, A., GOVAERTS, W., KUZNETSOV, Y.A., MESTROM, W., RIET, A.M., & SAUTOIS, B. 2006. MATCONT and CL_MATCONT: Continuation toolboxes in Matlab. *Gent University and Utrecht University*.

[40] DOEDEL, E.J., CHAMPNEYS, A.R., FAIRGRIEVE, T.F., KUZNETSOV, Y.A., SANDSTEDE, B., & WANG, X. AUTO 97: Continuation and bifurcation software for Ordinary Differential Equations (with HomCont).

[41] DOWSON, D., & TAYLOR, CM. 1979. Cavitation in bearings. *Annual Review of Fluid Mechanics*, **11**(1), 35–65.

[42] DOWSON, D., GODET, M., & TAYLOR, C.M. 1975. *Cavitation and related phenomena in lubrication*. Mechanical Engineering Publications Limited.

[43] DURAK, ERTUĞRUL. 2003. Experimental investigation of porous bearings under different lubricant and lubricating conditions. *Journal of Mechanical Science and Technology*, **17**, 1276–1286. 10.1007/BF02982469.

[44] DURAK, ERTUĞRUL, & DURAN, FAZLI. 2008. Tribological and fatigue failure properties of porous P/M bearing. *International Journal of Fatigue*, **30**(4), 745 – 755.

[45] ELROD, H.G. 1981. A Cavitation Algorithm. *ASME J. Lubr. Technol*, **103**(3), 350–354.

[46] ELROD, HG, & ADAMS, ML. 1975. A computer program for cavitation and starvation problems. 37.

[47] ELSHARKAWY, A.A. 2003. Effects of Misalignment on the Performance of Flexible Porous Journal Bearings. *Tribology Transactions*, **46**(1), 119–127.

[48] ELSHARKAWY, A.A., & GUEDOUAR, L.H. 2000. An Inverse Analysis for Steady-State Elastohydrodynamic Lubrication of One-Layered Journal Bearings. *Journal of Tribology*, **122**, 524.

[49] ELSHARKAWY, A.A., & GUEDOUAR, L.H. 2001a. An inverse solution for finite journal bearings lubricated with couple stress fluids. *Tribology International*, **34**(2), 107–118.

134

[50] ELSHARKAWY, A.A., & GUEDOUAR, L.H. 2001b. Direct and Inverse Solutions for Elastohydrodynamic Lubrication of Finite Porous Journal Bearings. *Journal of Tribology*, **123**, 276.

[51] ELSHARKAWY, A.A., & GUEDOUAR, L.H. 2001c. Hydrodynamic lubrication of porous journal bearings using a modified Brinkman-extended Darcy model. *Tribology International*, **34**(11), 767–777.

[52] FRÄNKEL, A. 1944. *Berechnung von zylindrischen Gleitlagern*. Ph.D. thesis, ETH-Zürich.

[53] GALASSI, M., DAVIES, J., THEILER, J., GOUGH, B., JUNGMAN, G., ALKEN, P., BOOTH, M., ROSSI, F., & PRICE, R. 2009. *GNU Scientific Library Reference Manual*. ISBN 0954612078.

[54] GASCH, R., NORDMANN, R., & PFÜTZNER, H. 2002. *Rotordynamik*. Springer.

[55] GIUDICELLI, B. 1993. *Modèles hydrodynamiques conservatifs pour des paliers autolubrifiants poreux en régime permanent*. Ph.D. thesis, Institut National des Sciences Appliqués de Lyon.

[56] GOENKA, P.K., MARTIN, F.A., ALLAIRE, P.E., BOOKER, J.F., & L.A. BOUFF, G.A. 1984. Dynamically loaded journal bearings: finite element method analysis. *Journal of tribology*, **106**(4), 429–437.

[57] GOLDSTEIN, M.E., & BRAUN, W.H. 1971. *Effect of Velocity Slip at a Porous Boundary on the Performance of an Incompressible Porous Bearing*. National Aeronautics and Space Administration.

[58] GREENWOOD, J.A., & TRIPP, J.H. 1970. The Contact of Two Nominally Flat Rough Surfaces. *Proc. Inst. Mech. Eng*, **185**, 625–633.

[59] GREENWOOD, J.A., & WILLIAMSON, J.B.P. 1966. Contact of nominally flat surfaces. *Proceedings of the Royal Society of London. Series A, Mathematical and Physical Sciences*, 300–319.

[60] GROSS, W.A. 1962. *Gas film lubrication*. Wiley New York.

[61] GROTE, K.H., & FELDHUSEN, J. 2007. *DUBBEL: Taschenbuch für den Maschinenbau*. Springer Verlag.

[62] GUHA, SK. 1986. Study of conical whirl instability of hydrodynamic porous oil journal bearings with tangential velocity slip. *Tribology international*, **19**(2), 72–78.

[63] GUHA, S.K., & CHATTOPADHYAY, A.K. 2007. On the linear stability analysis of finite-hydrodynamic porous journal bearings under coupled stress lubrication. *Proceedings of the Institution of Mechanical Engineers, Part J: Journal of Engineering Tribology*, **221**(7), 831–840.

[64] GÜMBEL, L.K.R. 1921. Vergleich der Ergebnissen der rechnerischen Behandlung des Lagerschmierungsproblems mit neuen Versuchsergebnissen. *Monatsbl. Berliner Bez. Ver. Dtsch. Ing*, 125–128.

[65] GURURAJAN, K., & PRAKASH, J. 1999. Surface roughness effects in infinitely long porous journal bearings. *Journal of Tribology*, **121**, 139.

[66] GURURAJAN, K., & PRAKASH, J. 2002. Roughness effects in a narrow porous journal bearing with arbitrary porous wall thickness. *International Journal of Mechanical Sciences*, **44**(5), 1003–1016.

[67] HAGEDORN, P. 1990. *Technische Mechanik, Band 3: Dynamik*. Harri Deutsch.

[68] HAQUE, R., & GUHA, S.K. 2005. On the steady-state performance of isotropically rough porous hydrodynamic journal bearings of finite width with slip-flow effect. *Proceedings of the Institution of Mechanical Engineers, Part C: Journal of Mechanical Engineering Science*, **219**(11), 1249–1267.

[69] HERREBRUGH, K. 1970. Elastohydrodynamic squeeze films between two cylinders in normal approach. *Transaction of the ASME, Journal of Lubrication Technology*, **92**, 292–302.

[70] HIGGINSON, G.R. 1966. The theoretical effects of deformations of the bearing liner on journal bearing performance. *Proceedings of the Institution of Mechanical Engineers, Part B: Journal of Engineering Manufacture*, **180**(3).

[71] HOLLIS, P., & TAYLOR, D.L. 1986. Hopf Bifurcation to Limit Cycles in Fluid Film Bearings. *Journal of Tribology*, **108**(2), 184–189.

[72] HUBBERT, M.K. 1956. Darcy's law and the field equations of the flow of underground fluids. *Trans. AIME*, **207**(7), 222–239.

[73] JAKOBSSON, B., & FLOBERG, L. 1957. The finite journal bearing considering vaporization. *Transactions of Chalmers University of Technology*, **190**.

[74] JANG, J.Y., & KHONSARI, M.M. 2010. On the Behavior of Misaligned Journal Bearings Based on Mass-Conservative Thermohydrodynamic Analysis. *Journal of Tribology*, **132**(1), 011702.

[75] JOHNSON, K.L. 1987. *Contact Mechanics*. Cambridge University Press.

[76] KANE, T.R., LIKINS, P.W., & LEVINSON, D.A. 1983. *Spacecraft dynamics*. McGraw-Hill Book.

[77] KANEKO, S. 1989. Static and Dynamic Characteristics of Porous Journal Bearings with Anisotropic Permeability. *JSME international journal. Series C, Mechanical systems, machine elements and manufacturing*, **32**(1), 91–99.

[78] KANEKO, S., & HASHIMOTO, Y. 1995. A Study of the Mechanism of Lubrication in Porous Journal Bearings: Effects of Dimensionless Oil-Feed Pressure on Frictional Characteristics. *Journal of Tribology*, **117**, 291.

[79] KANEKO, S., & OBARA, S. 1990. Experimental Investigation of Mechanism of Lubrication in Porous Journal Bearings: Part 1 - Observation of Oil Flow in Porous Matrix. *Journal of Tribology*, **112**, 618.

[80] KANEKO, S., OHKAWA, Y., & HASHIMOTO, Y. 1994a. A study of the mechanism of lubrication in porous journal bearings: effects of dimensionless oil-feed pressure on static characteristics under hydrodynalic lubrication conditions. *Journal of tribology*, **116**(3), 606–611.

[81] KANEKO, S., INOUE, H., & USHIO, K. 1994b. Experimental Study on Mechanism of Lubrication in Porous Journal Bearings: Oil Film Formed in Bearing Clearance. *JSME international journal. Series C, Mechanical systems, machine elements and manufacturing*, **37**.

[82] KANEKO, S., HASHIMOTO, Y., & HIROKI, I. 1997. Analysis of Oil-Film Pressure Distribution in Porous Journal Bearings Under Hydrodynamic Lubrication Conditions Using an Improved Boundary Condition. *Journal of Tribology*, **119**(1), 171–178.

[83] KANEKO, S., TAKABATAKE, H., & ITO, K. 1999. Numerical Analysis of Static Characteristics at Start of Operation in Porous Journal Bearings With Sealed Ends. *Journal of Tribology*, **121**, 62.

[84] KNOLL, G., & PEEKEN, H. 1990. FEM Formulation of Fluid Contact Elements for Elastohydrodynamic Structure Interaction. *In: Proc. of the Japan Int. Tribol. Conf., Nagoya, Japan.*

[85] KNOLL, G., SCHOENEN, R., & WILHELM, K. 1997. Full dynamic analysis of crankshaft and engine block with special respect to elastohydrodynamic bearing coupling. *In: ASME-ICED 1997 Spring Technical Conference, Fort Collins, Colorado.*

[86] KUMAR, A., & BOOKER, JF. 1991. A Finite Element Cavitation Algorithm. *Journal of Tribology*, **113**, 276.

[87] KUZNETSOV, Y.A. 1998. *Elements of applied bifurcation theory.* Springer New York.

[88] LAGEMANN, V. 2000. *Numerische Verfahren zur tribologischen Charakterisierung bearbeitungsbedingter rauher Oberflächen bei Mikrohydrodynamik und Mischreibung.* Ph.D. thesis, Universität Kassel.

[89] LAHA, SK, & KAKOTY, SK. 2011a. Effect of Unbalance on the Dynamic Response of a Flexible Rotor Supported on Porous Oil Journal Bearings. *Pages 229–239 of: IUTAM Symposium on Emerging Trends in Rotor Dynamics.* Springer.

[90] LAHA, S.K., & KAKOTY, S.K. 2011b. Non-linear Dynamic Analysis of a Flexible Rotor Supported on Porous Oil Journal Bearings. *Communications in Nonlinear Science and Numerical Simulation*, **16**, 1617–1631.

[91] LAWSON, C.L., HANSON, R.J., KINCAID, D.R., & KROGH, F.T. 1979. Basic linear algebra subprograms for Fortran usage. *ACM Transactions on Mathematical Software (TOMS)*, **5**(3), 308–323.

[92] LI, W.L., & CHU, H.M. 2004. Modified Reynolds equation for coupled stress fluids–a porous media model. *Acta Mechanica*, **171**(3), 189–202.

[93] LIN, JAW-REN, & HWANG, CHI-CHUAN. 1995a. Linear Stability Analysis of Short Porous Journal Bearings—Use of the Brinkman Model. *Journal of Tribology*, **117**(1), 199–202.

[94] LIN, JAW-REN, & HWANG, CHI-CHUAN. 1995b. On the Lubrication of Short Porous Journal Bearings—Use of the Brinkman Model. *Journal of Tribology*, **117**(1), 196–199.

[95] LIN, JAW-REN, HWANG, CHI-CHUAN, & YANG, RONG-FUH. 1996. Hydrodynamic lubrication of long, flexible, porous journal bearings using the Brinkman model. *Wear*, **198**(1-2), 156 – 164.

[96] LIN, J.R. 1995. Dynamic behaviour of pure squeeze films in short porous journal bearings using the Brinkman model. *Journal of Physics D: Applied Physics*, **28**, 2188–2196.

[97] LIN, J.R., & HWANG, C.C. 1993. Lubrication of short porous journal bearings: use of the Brinkman-extended Darcy model. *Wear*, **161**(1-2), 93–104.

[98] LIN, J.R., & HWANG, C.C. 1994. Static and dynamic characteristics of long porous journal bearings: use of the Brinkman-extended Darcy model. *Journal Of Physics - Applied Physics*, **27**, 634–634.

[99] LIN, J.R., & HWANG, C.C. 2002. Hopf bifurcation to a short porous journal-bearing system using the Brinkman model: weakly nonlinear stability. *Tribology International*, **35**(2), 75–84.

[100] LIN, J.R., LU, R.F., & YANG, C.B. 2001. Derivation of porous squeeze-film Reynolds equations using the Brinkman model and its application. *Journal of Physics D: Applied Physics*, **34**(22), 3217–3223.

[101] MAK, W., & CONWAY, HD. 1977. The lubrication of a long, porous, flexible journal bearing. *Trans. ASME J. Lubr. Technol*, **99**, 449–454.

[102] MAK, WC, & CONWAY, HD. 1978. Effects of velocity slip on the elastohydrodynamic lubrication of short porous journal bearings. *Int. J. Mech. Sci*, **20**, 767–775.

[103] MEURISSE, M.H., & GIUDICELLI, B. 1999. A 3D Conservative Model for Self-Lubricated Porous Journal Bearings in a Hydrodynamic Steady State. *Journal of Tribology*, **121**, 529.

[104] MOES, H. 2000. *Lubrication and beyond.*

[105] MOKHTAR, M.O.A., RAFAAT, M., & SHAWKI, G.S.A. 1984. *Experimental investigations into the performance of porous journal bearings.* TP-840097, Society of Automotive Engineers, Inc., Warrendale, PA.

[106] MORGAN, V.T., & CAMERON, A. 1957. Mechanism of lubrication in porous metal bearings. *Pages 151–7 of: Proc. Conf. on Lubrication and Wear.*

[107] MURTI, P.R.K. 1971. Hydrodynamic lubrication of long porous bearings. *Wear*, **18**, 449–460.

[108] MURTI, P.R.K. 1973. Effect of slip flow in narrow porous bearings. *ASME Journal of Lubrication Technology*, **95**, 518–523.

[109] MURTY, KG. 1974. Note on a Bard-type scheme for solving the complementarity problem. *Opsearch*, **11**(2-3), 123–130.

[110] NADUVINAMANI, NB, & MARALI, GB. 2007. Numerical Solution Of Couple Stress Full Reynolds Equation For Plane Inclined Porous Slider Bearings With Squeezing Effect. *Canadian Journal Of Pure And Applied Sciences*, **5**, 83.

[111] NADUVINAMANI, N.B., & PATIL, S.B. 2009. Numerical solution of finite modified Reynolds equation for couple stress squeeze film lubrication of porous journal bearings. *Computers & Structures*, **87**(21-22), 1287–1295.

[112] NADUVINAMANI, NB, HIREMATH, PS, & GURUBASAVARAJ, G. 2001. Squeeze film lubrication of a short porous journal bearing with couple stress fluids. *Tribology International*, **34**(11), 739–747.

[113] NADUVINAMANI, NB, HIREMATH, PS, & GURUBASAVARAJ, G. 2002. Surface roughness effects in a short porous journal bearing with a couple stress fluid. *Fluid Dynamics Research*, **31**(5-6), 333–354.

[114] NADUVINAMANI, N.B., SANTOSH, S., & SIDDANAGOUDA, A. 2010. On the squeeze film lubrication of rough short porous partial journal bearings with micropolar fluids. *Proceedings of the Institution of Mechanical Engineers, Part J: Journal of Engineering Tribology*, **224**(3), 249–257.

[115] NEACŞU, A., SCHEICHL, B., EDER, S., & VORLAUFER, G. 2011. Cavitation in porous journal bearings lubricated with ionic liquids. *In: Third European Conference on Tribology*.

[116] NEALE, G., & NADER, W. 1974. Practical significance of Brinkman's extension of Darcy's law: coupled parallel flows within a channel and a bounding porous medium. *Can. J. Chem. Eng*, **52**, 475–478.

[117] NIELD, D.A., & BEJAN, A. 1999. *Convection in Porous Media*. Springer.

[118] NIKOLAKOPOULOS, PG, & PAPADOPOULOS, CA. 1997. Dynamic Stability of Linear Misaligned Journal Bearings Via Lyapunov's Direct Method. *Tribology Transactions*, **40**(1), 138–146.

[119] OCVIRK, FW. 1952. Short-bearing approximation for full journal bearings. *Mem. National Advisory Committee for Aeronautics TN*, **2808**.

[120] OH, K.P., & HUEBNER, K.H. 1973. Solution of the elastohydrodynamic finite journal bearing problem. *ASME Journal of Lubrication Technology*, **95**(3), 342–352.

[121] OLSSON, K.O. 1965. Cavitation in dynamically loaded bearings. *Transactions of Chalmers University of Technology*, **308**.

[122] PARKER, T.S., CHUA, L.O., & PARKER, T.S. 1989. *Practical numerical algorithms for chaotic systems*. Springer New York.

[123] PATIR, N. 1978. *Effect of surface roughness on partial film lubrication using an average flow model based on numerical solution*. Ph.D. thesis, Northwestern University, Evanston, IL.

[124] PATIR, N., & CHENG, H.S. 1978. An average flow model for determining effects of three dimensional roughness on partial hydrodynamic lubrication'. *Transaction of the ASME, Journal of Lubrication Technology*, **100**(1), 12–17.

[125] PATIR, N., & CHENG, H.S. 1979. Application of Average Flow Model to Lubrication Between Rough Sliding Surfaces. **101**, 220–230.

[126] PEKLENIK, J. 1967. New developments in surface characterization and measurements by means of random process analysis. *Proceedings of the Institution of Mechanical Engineers*, **182**(311), 108–126.

[127] PINKUS, O., & BUPARA, SS. 1979. Analysis of misaligned grooved journal bearings. *ASME, Transactions, Journal of Lubrication Technology*, **101**, 503–509.

[128] POZRIKIDIS, C. 2009. *Fluid Dynamics: Theory, Computation and Numerical Simulation*. Springer.

[129] PRAKASH, J., & GURURAJAN, K. 1999. Effect of Velocity Slip in an Infinitely Long Rough Porous Journal Bearing. *Tribology Transactions*, **42**(3), 661–667.

[130] PRAKASH, J., & VIJ, S.K. 1972. Axially undefined porous journal bearing considering cavitation. *Wear*, **22**, 1–14.

[131] PRAKASH, J., & VIJ, S.K. 1974. Analysis of narrow porous journal bearing using Beavers–Joseph criterion of velocity slip. *Trans. ASME, J. Appl. Mechanics*, **41**, 348–354.

[132] RAMAN, R., & CHENNABASAVAN, TS. 1998. Experimental investigations of porous bearings under vertical sinusoidally fluctuating loads. *Tribology International*, **31**(6), 325–330.

[133] RAMAN, R., & CHENNABASAVAN, TS. 1999. Theoretical and Experimental Studies on Finite Porous Bearings with Tapered Journals. *Journal of the Institution Of Engineers India Part MC Mechanical Engineering Division*, **79**, 158–162.

[134] REASON, B.R., & SIEW, A.H. 1985. A refined numerical solution for the hydrodynamic lubrication of finite porous journal bearings. *Proceedings of the Institution of Mechanical Engineers. Part C. Mechanical engineering science*, **199**(2), 85–93.

[135] REYNOLDS, O. 1886. On the theory of lubrication and its application to Mr. Beauchamp Tower's experiments, including an experimental determination of the viscosity of olive oil. *Philosophical Transactions of the Royal Society of London*, **177**, 157–234.

[136] RIEMER, M., WAUER, J., & WEDIG, W. 1993. *Mathematische Methoden der Technischen Mechanik*. Springer.

[137] RITCHIE, G.S. 1975. The prediction of journal loci in dynamically loaded internal combustion engine bearings. *Wear*, **35**(2), 291–297.

[138] ROULEAU, W.T. 1963. Hydrodynamic Lubrication of Narrow Press-Fitted Porous Metal Bearings. *Journal of Basic Engineering*, 123–128.

[139] ROULEAU, W.T., & STEINER, L.I. 1974. Hydrodynamic Porous Journal Bearings, Part I: Finite Full Bearings. *Trans. ASME, J. Lubr. Technol*, **96**(5), 346–353.

[140] ROUTH, E.J. 1877. *A treatise on the stability of a given state of motion: particularly steady motion*. Macmillan and co.

[141] SEYDEL, R. 1996. BIFPACK: a program package for continuation, bifurcation, and stability analysis. *Program documentation, University of Ulm, Germany*.

[142] SHEA-BROWN, P. HOLMES, ERIC T. 2006. Stability. *Scholarpedia*, **1**(10), 1838.

142

[143] SOLOVYEV, S. 2006. *Reibungs- und Temperaturberechnung an Festkörper- und Mischreibungskontakten.* Ph.D. thesis, Universität Magdeburg.

[144] SOMMERFELD, A. 1904. Zur hydrodynamischen Theorie der Schmiermittelreibung. *Zeitschrift der Mathematischen Physik,* **50**, 97–155.

[145] STIEBER, W. 1933. Das Schwimmlager. *VDI, Berlin.*

[146] SWIFT, H.W. 1932. The stability of lubricating films in journal bearings. *Proc. Inst. Civil Engrs.(London),* **233**, 267–288.

[147] SZERI, A.Z. 1998. *Fluid Film Lubrication: Theory And Design.* Cambridge University Press.

[148] TANNER, R.I. 1961. The estimation of bearing loads using Galerkin's method. *International Journal of Mechanical Sciences,* **3**(1-2), 13–27.

[149] TOWER, B. 1883. first Report On Friction Experiments. *Proceedings of the Institution of Mechanical Engineers,* 632–659.

[150] TSCHENTKE, M. 2000. *Kriterien zur Auslegung hydrodynamischer Sintermetallgleitlager.* Ph.D. thesis, Universität Kassel.

[151] VAN BUUREN, S.W., HETZLER, H., HINTERKAUSEN, M., & SEEMANN, W. 2012. Novel approach to solve the dynamical porous journal bearing problem. *Tribology International,* **46**(1), 30 – 40. 37th Leeds-Lyon Symposium on Tribology Special issue: Tribology for Sustainability: Economic, Environmental, and Quality of Life.

[152] VAN DE VRANDE, B.L. 2001. *Nonlinear Dynamics of Elementary Rotor Systems with Compliant Plain Journal Bearings.* Ph.D. thesis, Technische Universiteit Eindhoven.

[153] VENNER, CH, & LUBRECHT, AA. 2000. *Multilevel methods in lubrication.* Elsevier.

[154] VIJAYARAGHAVAN, D., & KEITH JR., T.G. 1990a. An efficient, robust, and time accurate numerical scheme applied to a cavitation algorithm. *Journal of Tribology,* **112**, 44.

[155] VIJAYARAGHAVAN, D., & KEITH JR., T.G. 1990b. Analysis of a finite grooved misaligned journal bearing considering cavitation and starvation effects. *Journal of Tribology,* **112**, 60.

[156] WANG, Q.J., SHI, F., & LEE, S.C. 1997. A mixed-lubrication study of journal bearing conformal contacts. *Journal of Tribology,* **119**, 456.

[157] WIEGERT, B. 2008. *Nichtlineare Stabilitätsuntersuchungen am starren Rotoren mit Schwimmbuchsenlagerung.* Diploma Thesis, ITM, Karlsruhe Institute for Technology.

[158] WIJNANT, Y.H. 1998. *Contact Dynamics in the Field of Elastohydrodynamic Lubrication.* Ph.D. thesis, University of Twente.

[159] WILSON, W.R.D., & MARSAULT, N. 1998. Partial hydrodynamic lubrication with large fractional contact areas. *Journal of Tribology*, **120**, 16.

[160] WOODS, C.M., & BREWE, D.E. 1989. The solution of the Elrod algorithm for a dynamically loaded journal bearing using multigrid techniques. *Journal of tribology*, **111**(2), 302–308.

[161] YUNG, K.M., & CAMERON, A. 1979. Optical analysis of porous metal bearings. *Journal of Lubrication Technology*, **101**, 99–104.

KARLSRUHER INSTITUT FÜR TECHNOLOGIE (KIT)
SCHRIFTENREIHE DES INSTITUTS FÜR TECHNISCHE MECHANIK (ITM)

ISSN 1614-3914

Die Bände sind unter www.ksp.kit.edu als PDF frei verfügbar oder als Druckausgabe zu bestellen.

Band 1   **Marcus Simon**
Zur Stabilität dynamischer Systeme mit stochastischer
Anregung. 2004
ISBN 3-937300-13-9

Band 2   **Clemens Reitze**
Closed Loop, Entwicklungsplattform für mechatronische
Fahrdynamikregelsysteme. 2004
ISBN 3-937300-19-8

Band 3   **Martin Georg Cichon**
Zum Einfluß stochastischer Anregungen auf mechanische
Systeme. 2006
ISBN 3-86644-003-0

Band 4   **Rainer Keppler**
Zur Modellierung und Simulation von Mehrkörpersystemen
unter Berücksichtigung von Greifkontakt bei Robotern. 2007
ISBN 978-3-86644-092-0

Band 5   **Bernd Waltersberger**
Strukturdynamik mit ein- und zweiseitigen Bindungen
aufgrund reibungsbehafteter Kontakte. 2007
ISBN 978-3-86644-153-8

Band 6   **Rüdiger Benz**
Fahrzeugsimulation zur Zuverlässigkeitsabsicherung
von karosseriefesten Kfz-Komponenten. 2008
ISBN 978-3-86644-197-2

Band 7   **Pierre Barthels**
Zur Modellierung, dynamischen Simulation und
Schwingungsunterdrückung bei nichtglatten, zeitvarianten
Balkensystemen. 2008
ISBN 978-3-86644-217-7

KARLSRUHER INSTITUT FÜR TECHNOLOGIE (KIT)
SCHRIFTENREIHE DES INSTITUTS FÜR TECHNISCHE MECHANIK (ITM)

ISSN 1614-3914

KARLSRUHER INSTITUT FÜR TECHNOLOGIE (KIT)
SCHRIFTENREIHE DES INSTITUTS FÜR TECHNISCHE MECHANIK (ITM)

ISSN 1614-3914

Band 15   Aydin Boyaci
           Zum Stabilitäts- und Bifurkationsverhalten hochtouriger
           Rotoren in Gleitlagern. 2012
           ISBN 978-3-86644-780-6

Band 16   Rugerri Toni Liong
           Application of the cohesive zone model to the analysis
           of rotors with a transverse crack. 2012
           ISBN 978-3-86644-791-2

Band 17   Ulrich Bittner
           Strukturakustische Optimierung von Axialkolbeneinheiten.
           Modellbildung, Validierung und Topologieoptimierung. 2013
           ISBN 978-3-86644-938-1

Band 18   Alexander Karmazin
           Time-efficient Simulation of Surface-excited Guided
           Lamb Wave Propagation in Composites. 2013
           ISBN 978-3-86644-935-0

Band 19   Heike Vogt
           Zum Einfluss von Fahrzeug- und Straßenparametern
           auf die Ausbildung von Straßenunebenheiten. 2013
           ISBN 978-3-7315-0023-0

Band 20   Laurent Ineichen
           Konzeptvergleich zur Bekämpfung der Torsionsschwingungen
           im Antriebsstrang eines Kraftfahrzeugs. 2013
           ISBN 978-3-7315-0030-8

Band 21   Sietze van Buuren
           Modeling and simulation of porous journal bearings in
           multibody systems. 2013
           ISBN 978-3-7315-0084-1